Wild Silk Technology

Wild Silk Technology

R.P. Kavane
Department of Zoology
Vivekanand College
Kolhapur – 416 004
&
T.V. Sathe
Entomology Division
Department of Zoology
Shivaji University
Kolhapur – 416 004

2011
DAYA PUBLISHING HOUSE®
Delhi - 110 002

ISBN 81-7035-712-8
ISBN 978-81-7035-712-4

Published by : **Daya Publishing House®**
4760-61/23, Ansari Road, Darya Ganj,
New Delhi - 110 002
Phone: 23245578, 23244987
Fax: (011) 23260116
e-mail : dayabooks@vsnl.com
website : www.dayabooks.com

Laser Typesetting : **Classic Computer Services**
Delhi - 110 035

Printed at : **Chawla Offset Printers**
Delhi - 110 052

PRINTED IN INDIA

Dedicated to

SHIVAJI UNIVERSITY

Kolhapur

Preface

Sericulture plays a very important role in sustainable development of a country by providing valuable products. India is the second largest producer of the tasar silk in the world. Sericulture is practiced in diverse agro-climatic region across the length and breadth of the country which necessitates the evaluation of large number of region and season specific breed/hybrids. Non-mulberry sericulture known as forest or wild sericulture holds great promise for agro industry and forestry. Rearing of the wild silkworm specially, tasar silkworm provides employment and additional income to tribal families since traditional rearing method for cocoon production is less remunerative to the tribal rearers. Indoor rearing of tasar silkworms is possible up to 3rd instar but for 4th and 5th instars indoor rearing is yet to be established. In facts, rearing success rate of tasar silkworm is very less (30–35 per cent) due to number of predators, parasitoids and diseases. Hence, the present rearing technology provided in the text will be helpful for enhancement of wild sericulture activity in India. For better understanding of rearing concepts and utilization of tasar silkworm species in wild sericulture, there is urgent need of biosystematic studies of silkmoths. The present work will thus provide information of wild silkworms with respect to taxonomy, biology, and rearing potential with different food plants.

Hence, we hope that the book will be helpful to students, teachers, farmers, Sericulturists and researchers in the field of sericulture.

R.P. Kavane

T.V. Sathe

Contents

Preface *vii*

1. General Introduction 1
2. Review of Literature 37
3. Materials and Methods for Silk Technology 43
4. Taxonomy of Wild Silkmoths 55
5. Biology of Wild Silkmoths 121
6. Rearing Techniques for Wild Silkmoths 137
7. Summary 196
8. Conclusion 198

References 199

Index 221

Chapter 1
General Introduction

Silk not only refers to *Bombyx* silk whose insects are reared on mulberry leaves and is a high priced, valuable and delicate textile, called the "Queen of Fibres", but also to Tasar silks where tasar worms are reared on Ain, Oak, Arjun, Ber, Badam, etc. During the last century, the silk industry has greatly contributed to the foreign exchange to India. In the world, silk still offers the economic prosperity.

Mulberry silk is quite famous but other wild silks are not very popular, although Chinese tasar (*Antheraea pernyi* G.M), Indian tasar (*Antheraea mylitta* D), Eri silk (*Samia cynthia ricini* Hutt), Muga silk (*Antheraea assama* Ww) and Tensan (*Antheraea yamamai* G.M) have long been used for characteristic silk textiles, forming a small segment of the market. It has been observed through research that wild silks possess some physiological significance in activities like controlling cholesterol in blood, antibacterial functions and in UV absorption effect, etc. These newly recognized properties may greatly increase the popularity and utilization of wild silks at global scenario as world is facing the problems related to global warming. The International Society for Wild Silkmoths (ISWSM), Japan has started research work on wild silkmoths and silks. The society is established during 1980 and has used available technology to harness various means of wild silk utilization.

Based on porosity and the compactness wild cocoon filaments are classified into two types. The former has a porous structure in the cross-section of the filament, containing multiple sizes of fine tubules. The porous cocoon filament is characteristics of species of Saturniidae which includes *A. yamamai, A. pernyi, A. assama, A. mylitta, S.c ricini* and *Attacus atlas* L. The non-porous filament of cocoons is feature of all other families *viz.,* Bombycidae, Lasiocampidae, Thaumetopoeidae, and Psychidae. *Bombyx mori* L, *B. mandarina* M, *Gonometa postica* Walkar, *Anaphe panda* Bosid and *Cryptothelea formosicola* Strand, are very famous insects for this kind of silk from above families. Daba and Sukinda silkworms *A.mylitta* produce the thickest cocoon filaments with porous structure which is suitable for fashionable clothing like, sportswear. According to Akai (1998) porous filament maintains a constant temperature and humidity in a textile more than a compact filament. At the smallest class of porous filament, those of *Cricula trifenestrata* Helfer create a fine and soft fabric suggestive of cashmere.

The domesticated *Bombyx* silk has properties which affect cholesterol in hemolymph, alcohol metabolism, senile dementia and diabetes. In wild silks produced under environmental conditions, these factors are expected to be present even more strongly than in Bombyx silk (Akai, 1998). The antibacterial functions of *A. pernyi* and Eri silk are higher than that of *B. mori.* Similarly, the reduction of UV by *A. pernyi* is also superior.

The wild silks possess health related characteristics. Products made from these silks are popularized for these special qualities. Such added values of wild silks will stimulate their increased utilization at national and international level. Very recently, wild silk powder has been obtained (Akai, 1998). This powder too is expected to have health benefits and increase the demand for wild silks.

Until ten years ago, wild silk was related to only five species viz, *A. yamamai, A. pernyi, A. mylitta, A. assama* and *S. ricini* in the market. However, in recent years a few other wild silkmoth species have also came to front, *A. atlas, C. trifenestrata, A. panda* etc. but, there are yet no farms for these silkmoths and only cocoons collected from the field or hills are used. *Gonometa* and *Borocera* from family Lasiocampidae are also yielding interesting silks and have unique compact filaments. Establishment of the farms for these wild silkmoths is essential to introduce planned silk production in India. In the present work, *A. atlas, A. selene* and *A. mylitta indica* ssp nov have been tried for their rearing. The work will add great relevance to wild silk industry.

In India (Figure 1), Non-mulberry sericulture is an age old tradition, practiced mainly by the tribal people. When they do not have any work in agriculture and other allied pursuits non mulberry sericulture provides them moderate earnings in different lean seasons of the year. Wild sericulture remained obscure as an exclusive craft of tribal and hill folks inhabiting the forests of Central India, Sub-himalayan region and North-eastern India for long time. However, in recent years, this traditional craft of tribal has gained tremendous importance. Due to its rich production potential, eco-friendly nature of the activities and steady demand for hand made textile products within and outside the country etc, wild sericulture is commercially exploitated from traditional craft into an industry of high potential. As a industry it has an advantage of rich natural resources like food plants and tribal manpower. Utilising them to bring a balanced development without disturbing the existing ecological system is the great socioeconomic challenge in sericulture.

India is the only country bestowed by nature to produce all the four varieties of non-mulberry silks namely tropical tasar, oak tasar, eri and muga. Of the total raw silk production of 14035 MT during the years, 1997-98, the share of non-mulberry silk was only 2.3 per cent. From non-mulberry production, eri, tasar and muga silks account for 69.78 per cent, 24.85 per cent and 5.37 per cent, respectively (Shetty and Samson, 1998).

Host food plants for wild silkworms is crucial factor for wild silk industry. There is practically no systematic plantation of food plants for rearing non-mulberry silkworms in India. Therefore, Tasar silkworms are reared on food plants available in the forest. Muga silkworm rearings is conducted on food plants in forests/village grazing reserves etc and eri silkworm rearing is largely conducted indoors with the leaves of caster (*Ricinus communis* L) collected from nature grown trees. Recently, the research institutes of Central Silk Board (CSB) have introduced the concept of systematic plantation of food plants for all non-mulberry silkworms to improve the productivity of silks in India.

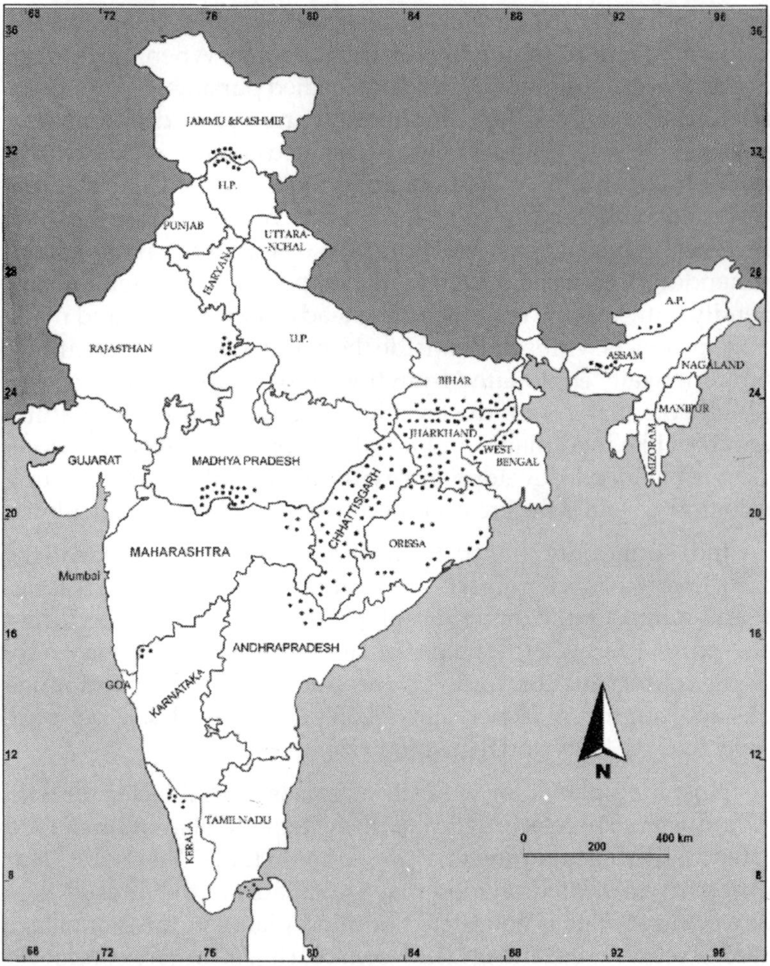

Figure 1: Map of India Showing Important States Cultivating Wild Silks

According to Sathe and Jadhav (2001) tasar silkworm rearing is practiced mainly in the Central and Southern Plateau region in the humid and dense forest area covering Bihar, Madhya Pradesh, Orissa and West Bengal, extending to the fringes of Uttar Pradesh, Andhra Pradesh and Maharashtra (Figure 1). It is estimated that in India, there is 11.168 million ha of forest having different primary and secondary food plants for wild silkworms being utilised for tasar silkworm rearing. However, deep interior forests remained unexploited. In India, about 1.40 lakh tribal families have been engaged in tasar silkworm rearing and get benefited socio-economically.

During 1981 and 1986 with the financial assistance from Swiss Development Co-operation in eight tasar producing states, the CSB has implemented the Inter State Tasar Project (ISTP), under this project, 7845 ha of *Terminalia arjuna* W&A block plantation has been developed with necessary infrastructural facilities and for overall development of tasar silk industry. Under ISTP Project, efforts have been made by Rajasthan Government for involvement of Rajasthan Vidyapeeth Kul, an NGO to take up tasar culture by utilising the plantation and other Infrastructure facilities.

Eri sericulture is practiced as an indoor activity. Eri silkworms are multivoltine and reared almost throughout the year. The ericulture has a close link with the culture and tradition of the people of North-east. Rural people are involved, primarily to meet the domestic demand of warm clothing and the edible pupae as a major source of proteins. The culture is also marginally practiced in West Bengal, Bihar and Orissa, primarily for production of castor seed and its oil. In general, eri culture is considered a subsidiary source of income for meeting the domestic needs of warm clothing and pupa. The rearers can produce their own seed and conduct rearing, spinning and weaving. The pupal dish of insect diet is sold in weekly market in eastern parts of India. Similarly, the surplus quantity of Eri cut cocoons are sold openly at their door steps.

Ericulture is competting with mulberry sericulture by all means. Muga, the golden yellow silk produced by muga silkworm is unique to Assam and neighboring states of North-eastern region and also practiced in West Bengal in recent years. Muga silkworm is multivoltine in nature with six crops a year, two each of commercial, pre-seed and seed crops. Muga silkworm is semi-domesticated and rearable in the open on trees where as spinning arid seed production are indoor activities. In 3,500 ha of land muga food plants are cultivated in North-eastern region of India including West Bengal.

The drawbacks of non-mulberry sector are poor performance of seed multiplication facilities and non-availability of adequate disease free commercial seeds. There is a great need to manage seed multiplications by the Government sector for quality seed and its timely supply to the rearers in India.

Recently, a four-tier seed multiplication system is established in India. The Central Tasar Research and Training Institute (CTR&TI), Ranchi is supplying the nucleus seed to Central Tasar Silkworm Seed Station, Lakha which multiplies the same for one generation and supplies to Basic Seed Multiplication and Training Centre (BMSTCs) for further multiplication. They in turn, multiply for one generation and supply the basic seed to Pilot Production Centre (PPC) of the state government. Each BSMTC produces 40,000 dfls and supply is 32000 dfls to eight PPCs @ 4000 dfls per PPC. Eight PPCs in turn produces 3.2 lakh dfls covering 3200 rearers @ 100 dfls per rearer.

In North-eastern and North-western states of India, Oak tasar culture has got good potential where abundant oak flora is available. However, Paucity of silkworm seed and lack of adequate extension support system are the main constraints for oak tasar development in this region. Therefore, a major emphasis should be given to this important basic need to ensure production and supply of quality basic silkworm seed in sufficient quantity to enhance the oak tasar silk production. CSB has established three research extension centres in order to fulfill the seed and extension requirements. These refer to REC in Assam, Nagaland and Manipur. CSB has also established the Regional Tasar Research Station (RTRS), at Bhimtal (UP) and RECS at Palampur (HP) in North-western region and these units produce and supply dfls to 35 seed farms in state sector for enhancing sericultural activities.

In India, all wild silkmoths *i.e., A. mylitta, A. assama, S.c ricini,* C. *trifenestrata, A. selene, A. atlas,* etc. and many more silkmoths distributed in diverse eco climatic zones. Similarly, considering the ecological conditions, food plant distribution, presence of eco-types and species of diverse nature is in coexistence. It is speculated that North-eastern India is a possible home of origin of species of *Antheraea* from where radiation would have occurred (Nagaraju and Reddy, 1998).

A. assama is supposed to be an ancestral species from which other species would have possibly originated. There are number of reasons to accept this concept. There are a few pointers to the hypothetical evolutionary course, all starting from the unique habitat, North-eastern India, one leading to Central and South India represented by the species of *A. mylitta.* The second leading to the North-West along sub-Himalayan belt represented by *A. roylei,* the third leading to Southern China and Japan represented by *A. pernyi* and *A. yamamai,* respectively and the fourth leading to Indo-Australian region represented by a number of tropical species. The only isolated case is that of *A. polypherous* distributed in USA. However, recent reports (Sathe, 2007) indicated that at least 15 species of wild silkmoths exists in Western Ghats of Maharashtra due to availability of rich flora of the region.

The Saturniid silkmoths have variations in chromosome number from n =15 to n = 49 with an average chromosome number of the family n = 31. Rare cases of interspecific hybrids of *Antheraea* shows diverse chromosome number of n = 31 and n = 49 are known to yield fertile and vigorous hybrids. According to Jolly *et al.* (1969) the interspecific hybrids generated using many other *Antheraea* species uncover interesting genetic relationships of the species involved. The holocentric nature of the chromosomes, presence of supernumerary chromosomes and chromosome numerical polymorphism in the species of *A. roylei* and fertility of the F_1 interspecific hybrids of *A.pernyi* and *A. roylei* despite trivalent formation show that the chromosomal fission has probably played central role in the evolution of Saturniid silk moths, (Puttaraju and Nagaraju, 1988).

Except *A. yamamai* in all the *Antheraea* moths where eggs undergo hibernation, the diapause sets in at pupal stage. Many species and eco-types have high degree of plasticity to shift from uni/ bivoltinism, trivoltinism depending on the photoperiod during larval stage and nutrition.

The biotechnological approaches such as recombinant DNA technologies, immunological methods, protoplast culture, hybridoma technology and gene transfer methods have provided solutions to many problems to conventional approaches in sericulture. The vast genetic resources from non-mulberry silk moths offer recent developments for expecting new rearable varities of wild silkmoths. In fact, the active research that is going on mulberry silkworm *B.mori* as a model genetic system has already provided important insights into the fundamental biological processess would worth implementing against wild silkmoths at global scenario.

A. mylitta as a polyphagous worm has fairly wide area of distribution in tropical belt extending from 16 to 24° N and 80 to 88° E covering both deciduous and semideciduous forests. It has 44 ecoraces distributed in India. The ecoraces such as Daba, Laria, Modal, Sukinda, Bhandara, Raily, Nalia, Sarihan and Andhra local are considered important from the commercial view point. These ecotypes differ sharply in their characteristics and offer rich genetic repertoire in terms of their qualitative and quantitative economic characters. Unfortunately, systematic use of these ecotypes is yet to be establish in tasar culture. Their diversity in voltinism, food plant preference and hybrid vigour manifestation has great practical potential value. The above study needs basic understanding of the genetic architecture of ecotypes, which would throw light on their uniqueness, genetic distances and genetic variability. Such information could be profitably used to identify the eco-types which give optimum level of heterosis in the hybrids, to preserve the genetic identity of the eco-types and to evolve breeding strategy for maintenance of genetic diversity (Puttaraju and Nagaraju, 1988).

Ecoraces have not studied with line of worth while economic interest. There is pronounced phenotypic and behavioral plasticity between ecoraces which would worth establishing their rearing relationships. The breeding programmes related to different eco-types conducted earlier could not succeed to the desired level. The reasons for such failures might be absence of well defined genetic markers. The breeding plans fail to reach the desired levels due to the fact of absence of genetic identity and genetic distance of the parents. Therefore, Tasar culture still waits for ways and means for rational characterization and utilization of ecotypes to arrive at high yielding, disease resistant and amenable hybrid combinations of silkworms. There are several recently introduced molecular marker technologies such as random amplified polymorphic DNA, inter-simple sequence repeat polymorphisms, simple sequence repeats based polymerase chain reaction, restriction fragment length polymorphisms etc. Such techniques are being intensively used in crop and animal genetics for protection of breeders' rights, conservation of genetic diversity, incorporation of desired traits through molecular marker assisted selection, germplasm characterization, genome mapping and map based cloning (Shi *et al.*, 1995; Nagaraja and Nagaraju, 1995; Nagaraju *et al.*, 1995).

The fertile interspecific hybridization is a rare phenomenon in insects. Genetic analysis of such fertile interspecific hybrids will throw light on the phylogenitic relationships of the wild silkworm species. Jolly *et al.* (1969) reported the cases of fully and partially fertile interspecific hybrids in the genus *Antheraea*. Nagaraju and Jolly (1986) studied the species of *A. pernyi* (Chinese oak silk moth) (n = 49) and *A. roylei* (Indian silk moth) (n = 30, 31), followed by the studies of Kobayashi and Tanaka (1988) and Shimada and Kobayashi (1992) involving *A. yamamai* (n = 31) and *A. pernyi* (n=49).The hybrids have quite interesting features in one full and the other partial fertile interspecific. The interspecific hybrid involving *A. pernyi* (n = 49) and *A. roylei* (n = 31) produces fully fertile Fl hybrids and could be inbred and backcrossed as well to *A. pernyi*. The chromosome pairing pattern in the fertile F_1 hybrid revealed 18 trivalent and 13 bivalents. Wherein stabilized chromosome number of n = 49 in the interspecific hybrid population inbred was found for 49 generations. It is very likely that the chromosomes which involved in trivalent formation were gradually excluded during in breeding and only those zygotic combinations which contain bivalents would have been retained in the population. Thus, the present day inbred hybrid of *A. pernyi* x *A. roylei*, referred to as *A. proylei* has only n = 49(Nagaraju, 1998). However, the fact is unknown as to what extent the parental genomes are represented in the stabilized inbred populations of *A. pernyi* and *A. roylei* (*A. proylei*). According to some workers many silk yield attributes and the vigour have substantially declined in the present inbred populations of *A. proylei*. The molecular genetic analysis of the interspecific hybrid and descendant populations, using DNA markers and chromosome painting, probes could explain the status of *A. roylei* chromosomes, which involved in the trivalents during the course of interbreeding of the hybrid.

The work on retaining and breeding a large number of inbred derivatives from backcross and F_2 populations of the interspecific hybrid, which are likely to carry a different of chromosomal complements to track the parental chromosome, by using mapped DNA markers and chromosome painting probes have great importance in wild sericulture development.

According to Kobayashi *et al.* (1982) there are conflicting observations with regard to the fertile nature of the *A. yamamai* and *A. pernyi* interspecific hybrids. Shimada and Kobayashi (1992) have not been able to advance the interspecific hybrid beyond F_1 generation. But, Kobayashi *et al.* (1982) have advanced the interspecific hybrid to F_2 generation. Shimada and Kobayashi (1992) have successfully obtained progenies from the backcross of F_1 male (*A. yamamai* female X *A. pernyi* male) X female of *A.yamamai*. Such success opens up the possibilities of raising new breeds that combine the high quality silk of *A.yamamai* and bivoltinism of *A. pernyi*.

Biotechnological advances have made it possible to express foreign genes in heterologous organisms. The main goal of such work should be to develop methods for efficient production of a large amount of purified, biologically active proteins to study the basic mechanism of gene expression and the biological effects of products in cells and organisms and their possible pharmaceutical use. Baculo viruses are extremely useful vectors for their successful expression of biologically active proteins. For expressing proteins of biomedical importance, a good progress have been made in *B. mori* nuclear polyhedrosis virus (BmNpV). Zhang *et al.* (1992) developed the recombinant expression vector of *A. pernyi* nuclear polyhedrosis virus (ApNPV) successfully and used for the expression of DE protein in *A. pernyi* pupae. *A. pernyi* can be considered as an ideal host to express the foreign proteins using ApNPV due to its advantages such as pupal diapause when pupa sleeps for several months, its sensitivity to infection and cheap cost of production.

Insects are capable of detecting foreign cells and foreign molecules of eliciting effective and specific defensive response. Among the components of immune response, wound healing and haemocytes and synthesis of a battery of defense proteins have been investigated. The synthesis of bacteria elicited proteins has been reported in *A. pernyi* (Qu *et al.*, 1982), *Philosamia cynthia* D (Boman and Hultmark, 1987) and *A. mylitta* (Nagaraju *et al.*, 1992). But, the novel antibacterial proteins, their regulation of expression and their precise role in immune response in Saturniid silk moths is to be detected in near future.

Silk proteins, particularly, fibroin, small fibroin, and P_{25} which constitute actual silk fibre are produced only in the posterior silk gland and sericin protein which cements the silk fibre is produced only in the middle silk gland in a highly tissue specific manner is demonstrated in *B.mori*. The genes coding for all these proteins have been cloned and characterized. However, *Bombyx* and *Yamamai* fibroin genes are quite different in their structure but their regulatory regions are conserved.

According to Kundu *et al.* (Personal communication) and Syed (personal communication) in *A. mylitta* and *A. assama* the amino acid sequence and composition are quite different from *Bombyx*. Silkmoths are holometabolous insects and they manifest robust circadian behaviours that involve in the timing of the photoperiodic termination of pupal diapause, adult eclosion and egg hatching behaviour.

The discovery of period gene that codes for period protein revealed that period gene expresses early in four cells of each of brain hemispher, with one lateral pair and one medial pair in the dorsolateral region (Sauman and Rappert, 1996). The period gene has important role in the circadian system. Extension of such studies to Saturniid silk moths on circadian behavioural pattern would be helpful for managing wild silk in better way in sericulture business.

Wild silkworms exhibit well defined taxonomical diversity. Their biodiversity is well known in all aspects of their life from egg to moth. Diversity in food habit and physiological constitution has been illustrated in *A. mylitta* (Indian tasar) *A. pernyi* (Oak tasar), *A. roylei* (Wild oak tasar), *A. Assama* (Muga), *Attacus synthia* (Wild eri) and *A. ricini.*

The tasar silkworm *A. mylitta* in Central India is being fed on Sal *Shorea robusta* Roxb and its "Daba" variety in northern Bihar (India) feeds on Asan *Terminalia tomentosa* W&A and *T. arjuna*. Attempts to rear Raily, the ecorace of tasar on *T. tomentosa* and *T. arjuna* resulted in high mortality with poor specimens of cocoons. While, Sal based raily are characterized by black coloured short peduncle and deep brownish grey to blackish cocoons. They are hard as stones, rich in silk content and the silk thread is thick with high denier. The raily silkworms are predominantly green with or without lateral shining spots. They camouflage perfectly with the back ground of Sal leaves. The race goes in diapause in pupal stage and complete one life cycle in a year showing univoltinism. The male moths are dark brown to deep brick red colour and females are deep yellow to brownish grey. Males have wings well adopted for long flights before mating. The eggs are different from Daba variety by a specific follicular imprint.

The primary food plants of Daba ecorace, *T. tomentosa* and *T. arjuna* are grown in Southern Bihar and bordering Orissa state. However, rearing these silkworms on Sal is not successful. Daba tasar silkworms are adopted to the warmer and humid climate in the months of June–July-August as a seed crop while, without diapause, second generation is adapted to less humid and comparatively favorable cooler seasons of September-October-November as commercial crop exhibiting bivoltinism. However, seed crop is poor in structure and silk content, the 2nd crop yields hard cocoons with rich silk content. In November–December the pupae of 2nd crop undergo diapause and emerge as moths in July next year.

According to Narasimhanna (1998) the Daba ecorace has shown tremendous potentiality to adopt and survive adverse and favorable climatic factors with the ecosystem unlike the univoltine raily, Berharva and Bhandara ecoraces of tasar. The egg structure, larval pattern and moth characters are morphologically different in Daba than that of raily. The larvae also show following differences that they exhibit fast colours–green, yellow, blue and whitish with adoptability for camouflage in the new environment for protection of itself from different types of predators. The peduncle of Daba cocoon is longer than raily with dark-brownish to yellowish coloured cocoons. Sal based cocoons are darker in shade. Daba cocoons are comparatively easy to rear and the silk produced is of thinner denier than that of raily. Moths are lighter coloured in Daba than those of raily, Bhandara and Berharva ecoraces. Daba moths are vulnerable to breeding in captivity unlike their counterpart of central India.

A. mylitta can also shows biodiversity in the form of Sarihan ecorace of eastern part of Bihar, Bogai ecorace of Orissa, Laria ecorace of southern Bihar (Ranchi) and Andhra ecorace of Adilabad in Andhra Pradesh. In all these ecoraces, three generations are completed in a single year. During the first two generations, they struggle for existence in adverse conditions and only those which are fit, survive to produce commercial cocoons during favourable seasons. Andhra ecorace is adopted to *T. tomentosa* which is available as shrub unlike the big trees of South Bihar where Daba ecorace is grown. The larvae are characterized by having smaller in size and

Table 1: Ecoraces of Indian Wild Silkmoths

Ecoraces	*State*	*Cocoon Weight (g)*	*Shell Weight (g)*	*SR (%)*	*Peduncle Length (cm)*	*Cocoon Size*		*Predominant Cocoon Colour*	*Primary Food Plant*
						Length (cm)	*Width (cm)*		
Tropical Tasar (*Antheraea mylitta*)									
Andhra Local	Andhra Pradesh	8.14	1.37	16.93	3.6	3.0	2.5	Whitish grey	*Termlnalia* sps.
Nowgaon	Assam	8.15	0.94	11.82	3.5	3.8	2.9	Grey	*Terminalia* sps.
Barharwa	Bihar	10:35	1.90	18.82	5.8	4.8	3.4	Grey	*Shorea robusta*
Daba	Bihar	11.95	1.79	16.06	6.8	5.7	3.7	Grey	*Terminalia* sps.
Laria-P	Bihar	7.78	1.63	20.97	5.2	5.0	3.1	Blackish grey	*Shorea robusta*
Moddia	Bihar	12.91	2.84	22.25	3.2	5.7	3.5	Grey	*Shorea robusta*
Mugia	Bihar	8.81	1.93	22.05	6.5	5.2	3.8	Grey	*Shorea robusta*
Munga	Bihar	5.98	0.76	12.79	5.8	4.8	3.3	Grey	*Terminalia* sps.
Sari ha n	Bihar	7.30	1.03	14.08	4.1	3.6	2.6	Grey	*Terminalia* sps.
Raily-N	Madhya Pradesh	11.86	2.15	18.44	3.5	5.3	3.4	Blackish grey	*Shorea robusta*
Bhandara	Maharashtra	7.30	1.51	20.76	4.5	4.1	2.4	Grey	*Terminalia* sps.
Giribum	Manipur	9.82	1.65	16.85	3.4	2.9	2.2	Whitish grey	*Zizyphus auritiana*

Contd...

Table 1–Contd...

Ecoraces	*State*	*Cocoon Weight (g)*	*Shell Weight (g)*	*SR (%)*	*Peduncle Length (cm)*	*Cocoon Size*		*Predominant Cocoon Colour*	*Primary Food Plant*
						Length (cm)	*Width (cm)*		
Bogai	Orissa	7.61	1.33	18.04	6.4	4.3	3.8	Whitish grey	*Terminalia* sps.
Modal	Orissa	13.61	2.98	22.12	6.4	5.4	3.5	Blackish grey	*Shorea robusta*
Nalia	Orissa	11.8	2.14	18.10	7.2	4.1	2.4	Grey	*Shorea robusta*
Sukinda	Orissa	11.22	1.71	14.49	5.3	4.9	3.1	Yellow	*Terminalia* sps.
Sukly	Orissa	7.38	1.16	15.83	3.4	3.7	2.8	Whitish grey	*Terminalia* sps.
Moonga	Uttar Pradesh	6.49	1.10	17.07	5.8	4.5	3.7	Grey	*Terminalia* sps.
Tiira	West Bengal	5.82	0.92	15.86	4.6	2.5	2.3	Whitish grey	*Lagerstromea* sps.

green colouration mostly. But, occurance of four colours green, yellow, blue and white is reposed in Sarihan, probably due to adoptation to a particular eco factor. Nature of cocoons produced are also different.

Laria ecorace is characterized by a long peduncle and yellow cocoon, while the cocoons of Bogai and Sarihan ecoraces are small in size with deep brown to black peduncles. However, moths are smaller in size than Daba ecorace with predominantly brick red males and yellow and greyish Females. The tasar silkworm *A. mylitta* thus, shows biodiversity in morphological, physiological and economic features. Sal based raily ecorace is univoltine. Daba is bivoltine and Sarihan is trivoltine. Daba and Sarihan are adopted themselves to the variable seasonal climatic conditions of ecological areas. Interbreeds between these races have been tried but, the attempts were not so successful for commercial exploitation leading to a question whether the status of such species can be confirmed or ecoraces to be retained for these tasar silkworms of different regions.

Attempts have been made in the crossbreed of *A. pernyi* and *A. roylei.*

The above species feed on oak plants. *A. roylei* is wild and *A. pernyi* is a cultivated species. *A. pernyi* produces fine reelable, shining silk and the cocoons of A. *roylei* can not be reeled or spun, hence of little economic importance to that of *B. mori* and *A. yammamai*. In north and north eastern parts of India *A. pernyi* is commercially exploited for its silk as it is well adopted to colder and warmer eco systems with their monocycle and bicycle life style. *A. pernyi* produces brownish cocoons with thin peduncle and single layered, while cocoons of *A. roylei* are greenish with two layered and multi cornered outer shell. The larvae also differ morphologically. Similarly, the moths are also different as reddish brown in *A. pernyi* and greenish grey in *A. roylei*. Their hybrids are also viable.

The Japanese tasar silkworm *A. yammamai* is confined to central Japan. Silk of *A. yammamai* is characterized by lustrous cocoons with fine denier, making it a most sought silk specially for embroidery. The silk *A. yammamai* is having great demand.

Muga silkworm, *A. assama* is confined to Brahmaputra valley of Assam, in India which feed on Som and Soalu plants. *A. pernyi* and *A. mylitta* can also feed on above plants. The seeds of this species are available for rearers and is commercial cocoon crop. However, the mortality of larvae is high in seed crop leading to shortage of eggs for commercial purpose. Thus, it is belived that the species is ill adopted to the seed cocoon season.

The tasar silkworms continue to feed on the leaves of host trees, but, *A.* assama larvae crawl down after the leaves are consumed. Those worms are picked and transfered to other plants. Mature larvae also crawl down which are picked and put to baskets containing leaves for spinning cocoons. This is unique habit of the silkworm found in *A. assama* which is not seen in any of the wild silkworms. The cocoons of *A. assama* are compact, single layered and without peduncle. The brownish coloured cocoons can be reeled on special contravences. The silk is brilliantly lustrous brownish in colour. The species exhibits uni, bi and trivoltinism in wild state. No brilliant colours are seen in the moths as seen in *A. mylitta*. Thus, the moths are dull reddish brown coloured in *A. assama*.

Next to mulberry silkworm *Bombyx mori* L., Eri silkworm *Philosamia ricini* is the only wild silkworm reared indoors. Their regular host plant is castor (*R. communis*) but, occasionally they are reared on Kesseru (*Heteropanax fragrans* Seem) leaves. These worms are well adapted to high humid ecofactors of Assam region. The hardy larvae exhibit green, blue, yellow and white colours like those of *A. mylitta* with difference in tubercle structure. *P. ricini* spin white coloured soft cocoons while, the wild *A. cynthia* spin brick red coloured soft cocoons. The cocoon structure of these is different from any other wild silkworms. However, the cocoons of *A. mylitta* are hard, with thick peduncle, and those of *A. pernyi* are comparatively soft and thin peduncled. In *A. ricini* the outer layer of cocoon is hairy, soft and the inner layer is comparatively harder. These cocoons are entirely different from the cocoons of *A. roylei* wherein outer layer is harder and inner layer softer. The moths are black in colour and do not exhibit bright colours of *A. mylitta*. The cocoons are soft and not reelable because of the ununiform of fibre. They are fit only for spinning. The silk of *A. ricini* poor quality and lacks luster. The rearing of this species is more related to human diet (pupae are eaten by Tribals) than for the silk. These Cocoons are sold after removal of pupae or after emergence of moths. Eggs are produced indoors and are laid on Khorikas (Narasimhanna, 1998).

Biodiversity within a species like tasar reveals its potentiality and the genetic adaptability through interaction with environment, to struggle, survive in varying ecological nitches. Biodiversity is a way of life in wild silkworms and well adopting to the living conditions in adverse ecological situations for yielding valuable silk for mankind.

The availability of food plants for non-mulberry silkworms has tremendous importance in enhancement of the non mulberry silks. Therefore, making the index of non mulberry silkworms and their host plants is crucial task in wild sericulture. The tropical tasar have been endowed by nature with vast forests food plants but, temperate tasar have different species of *Quercus* distributed throughout the temperate zone of north hemisphere and extended up to the tropics and subtropics of India, as food plants. For muga silkworm (*A. assamensis*), Som (*Machilus bombycina* King) and Soalu (*Litsea polyantha* Juss) are the two principal food plants. For eri silkworm Castor (*R. communis*), is main food plant and Kesseru (*Heteropanax fragrans* Seem) and Tapioca (*Manihot utilissima* Pohl), are secondary food plants. However, the non-mulberry food plants are not available in organized form. The tropical tasar food plants *T. arjuna* (Arjuna), *T. tomentosa* (Asan), Oak, Som and Soalu are the forest plants while Castor, Kesseru and Tapioca plants of eri silkworms are found often in the gardens/backyards of farmers, due to their manifold uses (Sinha 1998). Therefore, from sericulture point of view, a large population of food plants is required in an organized form which may be cared suitably to provide quality leaves for silkworm feeding. Naturally, they are well distributed in different states of the country and occupy 25 per cent area of the total forest cover. In states of Madhya Pradesh, Bihar and Orissa, the wild silkworm food plants are widely distributed. Some tasar food plants do exist in the villages on bunds of fields which farmers primarily use as shade trees can be managed for sericulture purpose.

The Western-Himalayan range at 1200 mtrs and in the eastern hilly tracts which constituting a major part of the forest are good areas for different species of oak in India. Traditionally, muga culture is practiced on irregularly scattered nature grown tall trees along the sub-himalayan hill ranges, particularly in the North-eastern India.

Regular cultivation of castor is done for oil seed and other purposes than sericulture in North-eastern states including Assam, the homeland of ericulture. Same is true with tapioca in Southern states are potential areas of ericulture in India. However, ericulture is now recommended by CSB for almost every states of India because, where caster is grown, ericulture is possible

The ever increasing urbanization and lopping of forest have led to gene erosion of many valuable genotypes of various species of non-mulberry flora (Sinha 1998). Therefore, suitable strategies for conservation of genes and genotypes of these sericultural flora and fauna to maintain bio-diversity and to continue the age-old tradition of non-mulberry sericulture required even at high premium. According to New Forest Policy (1952) the area under forests should be upto 33.3 per cent of the country's total land area. Government of India adopted this resolution. The forests which have protective role in balancing the environment have been depleting in great proportion. Even the tribals are also adopted for lopping of forest trees for their livelihood. In the absence of any alternative forest, same policy forests may shrink at a faster rate. Non- mulberry sericulture is one of the most important alternative means of livelihood of 10 per cent tribal population of India. Therefore, wild sericulture can save forests and simultaneously improve the socio-economic status of tribals and helps to promote suitable use of forest and safeguards their genetic resources. Soil erosion and several environmentally problems are related to wild sericulture. Wild sericulture thus, adds great relevance in solving environmental problems.

According to Khosla (1988) the concept of genetic conservation has developed with rapidly narrowing genetic base of traditional varieties. Though a large number of proposals for conserving gene resources in forest trees have been made, only a few of them are being practiced for non mulberry food plants. There are basically two types of conservation strategies *viz. in situ* conservation in natural strands and *ex situ* conservation both in living conditions and storage banks.

So far, no genetic resources of non-mulberry food plants have been established, hence attempts should be made in this direction in every state for their inhabitant species by the concerned forest department or sericulture department. Seed strands or seed production areas should be established in different states which can serve as control or check lots. Tree selection is also one of the methods to conserve the maximum diversity at species level and can be preserved through cloning. Attempts for cloning in most of the non mulberry food plants have been made but with some limitations propagation through cloning of perennial food plants of eri are yet to be established.

Knowledge of breeding system, biology and biological characters of the species is the pre-requisite for ex situ conservation the ex situ conservation is expensive to establish and maintain and therefore, they are confined to species of proven potential value. Seed orchards of non-mulberry food plants should be established primarily for the production of seeds of genetically proven quality. There are two types of seed orchards, clonal and seedling. They can be used as germplasm material for mass multiplication by vegetative means. The non-mulberry food plants parts viz, seed, fruit, pollen and tissue should be exploited on large scale in ex situ conservation of germplasm of flora of wild sericulture.

The tasar culture is largely confined to the forests due to the existence of a fairly large amount of forest flora to required for the culture. There is need to systematic plantation of good varities of non mulberry food plants due to lack of systematic plantation in North-eastern India has restricted the muga culture to a limited pocket only and same is the case of eri culture. Through abundant flora is available, in the region, non-mulberry sector has proved non-productive due to the above reseasons.

The tropical tasar of *A. mylitta* which is reared on *T. arjuna* and *T. tomentosa.* Though a large quantity of tasar cocoons is collected from Sal (S. *robusta*) forests, it has never been considered as a dependable food plant of *A. mylitta.* Central Tasar Research and Training Institute (CTR&TI), Ranchi developed widely acclaimed technology for multiplication of these food plants. The ripe and mature seeds are collected during March and April. Then grading is given according to maturity. Seeds are soaked for 196 hrs in case of *T. arjuna* and 48 hrs in *T. tomentosa.* The soaking is done in plain water. The soaked seeds are heaped under tree shade and covered with moist gunny bags. After six days the seeds start germination. The germinated seeds are dibbled on polythene tubes of 25 × 10 cm.

size which contain rooting mixture of FYM, soil and sand (3:2:1). Regular watering and care is taken in nursery beds and such seedlings are ready for field transplantation with three months. 30 × 30 × 30 cm size pits are used for transplantation with spacing of 1.25 × 1.25 m in case of *T. arjuna* and at 1.80 × 1.80 m in case of *T. tomentosa*.

Mainly two distinct silk varieties mulberry and non-mulberry are found in the world. Non-mulberry silk refers to Tasar, Muga, Eri, Anaphe, Fagara, Sinew, Mussel, Spider and Coan. India has a distinction of producing more number of commercial varieties than the word. Tasar, Muga and Eri are well known silks of India. *A.mylitta* for tasar silk is grown exclusively on *Terminalia, Shorea* and *Zizyphus* and known by its Indian name Tasar. The golden yellow silk of Assam produced by *A. assama* is known as Muga which is grown on Som (*M. bombycina*) and Soalu (*L. polyantha*). The brick red or white Eri silk is produced by *Philosamia ricini* Bosid. The silkworms of this species feed on castor leaves (*R. communis*).

Problems Related to Wild Silk

The wild silk culture have metrological and technical problems of its own. Unlike mulberry, tasar and muga silkworms are mostly reared outdoors and exposed to the vagaries of nature. Therefore, the wild silk culture has uncertainties in respect of viability, disease free layings to cocoon ratio and *vice-versa*. Besides, it has enormous variability through its various stages from egg to moth as also the commercial characters making any breeding programme cumbersome.

Problems Related to Tasar

There are two distinct varieties tasar *viz.*, tropical and temperate. The tropical tasar silkworm *A. mylitta* is cultivated by aboriginals of Bihar, Madhya Pradesh, Orissa, West Bengal, Uttar Pradesh, Maharashtra and Andhra Pradesh. The temperate tasar silkworm *A. proylei* is cultivated on *Quercus* species in the North-East and North-West sub-himalayan belt of the country. Both the varieties have distinct technical problems of their own.

1. Unorganized tasar food plants and their depletion due to gradual deforestation.
2. Due to outdoor rearing especially on tall trees, the early stage worms are exposed to vagaries of climate, pests and

predators causing about 20-30 per cent mortalities in larval population.

3. Unseasonal erratic emergence of moths causes loss to seed stock.
4. Poor fecundity/hatching and tasar silkworm diseases specially, pebrine bacteriosis, viruses and mycosis affect tasar culture adversely.
5. Traditional cooking and reeling processes are less productive and less cost effective.

Problems of Oak Tasar

1. Crop instability due to fast maturity of oak foliage (*Q. incana/Q. himalayana*) in North-western sector.
2. Pairing inefficiency, low fecundity (100-150 eggs)/ hatching.
3. Poor yield of cocoons (10) per disease free layings and thereby low productivity. Uzi fly menace decreased productivity in North-Eastern sector since rearing is outdoors.
4. Unseasonal erratic emergence of moths due to weak voltinism leads to loss of seed stock.
5. Due to lack of proper reeling technology the silk recovery is very poor.
6. No region specific subspecies/ strains/races/varieties are visualized.

Development in Tropical Tasar

CSB and CSR&TI, Ranchi have developed following strategies for improving tasar sericulture:

Nursery Technique for Raising Arjun and Asan Seedlings

For raising Arjun (*T. arjuna*) and Asan (*T. tomentosa*) seedlings in a large scale nursery techniques were evolved which ensured uniform and quick seed germination of 85-90 per cent and 60 per cent respectively. The successful raising of seedlings of nature grown Arjun and Asan trees helped in the implementation of inter State Tasar Project leading to about 8000 ha of Arjun bush plantation in tasar growing states. This is a major step taken towards augmentation of host plant in Tasar sector.

Introduction of Scheme of Economic Plantation

Economic tasar bush plantation with 4' × 4' spacing was found to be ideal for maximum leaf yield. Therefore, economic plantation in one ha land accommodates 6,724 plants and the yield of leaf at the end of 4 years is 18 MT. The technique has been demonstrated to generate interest in the rearers for its adoption.

Good Strategy for Pest Management of Host Plants

A suitable technique has been developed to minimize the damage to host plants from the gall insect (*Trioza fletcheri* Minor), stem borer (*Sphenoptera* sps.) and defoliating insect (*Notolophus* sp.) which are the major pests of tasar food plants. For minimising attack of parasites and predators to tasar silkworm, an integrated pest management strategy has been developed which involved judicious use of chemicals and biological control agents.

Package to Silkworm Diseases Control

The Package of practices both for preventive and curative measures against diseases of tasar silkworm *A. mylitta* specially, pebrine, bacteriosis, viruses and mycosis have been suggested. The diseases causes 30-40 per cent loss to sericulture industry. Tasar Keet Oushadh (TKO) has been formulated to control the bacteriosis, viruses and mycosis.

Improvement in the Rearing Techniques

The controlled rearing technique has been evolved for first stage worms in indoor conditions. Further, to increase yield and quality of the foliage during rearing period, foliage spray of 1.5 per cent urea on *T. arjuna* improved the yield and quality of leaves. At the same time, yield of cocoons per dfl increased to 70-80 cocoons which against previously had 20 to 25 under traditional methods.

Isolation of Trivoltine Races

A stable trivoltine line of Daba and Sukinda races has been isolated successfully which helped in the introduction of third crop rearing with higher ERR.

Seed Preservation Technique

An ideal grainage house with mud wall; thatched or tiled roof of 15' × 20' dimension with 8' all round verandah has been recommended for preservation of one lakh cocoons. The grainage has been designed keeping in view the higher temperature conditions prevailing in tasar producing states, during summer months.

Improved Technologies for Silk Extraction

For softening and making cocoons efficiently reelable, Proteolytic enzymes *viz.*, cocoonase, Biopril-50, Papain, Trypsin and Pepsin have been screened. Raw papaya has been also introduced as a cocoon softening agent with least damage of silk quality. While, 100 per cent cooking is ensured by pressure cooking.

Fabrication of Reeling Machine

For producing 400 gm of reeled silk in 8 hr a four spindle reeling machine has been fabricated and demonstrated. Therefore, traditional techniques can be replaced. Further, studies are in progress to reel the pierced/ emerged cocoons which are hitherto considered unreelable. It has been observed that the reelings of pierced/emerged cocoons are more profitable than using the ghicha technique. The above improvement in the technology will certainly encourage the wild sericulture in India.

Recently, a technique has been improved to the reeling and spinning of cocoons through sustainable research and technology upgradation programmes by the Central Silk Board. The traditional technique was crude and primitive, for preparation of silk yarn from the wild silk cocoons.

Reeling of Tasar Cocoons

The cocoons of tasar are hard, they are cooked in water with a pinch of soap and soda for 5-6 hours. The cooked semi-dried cocoons are deflossed by hand to get the filament ends. Five to six cocoon filaments are combined together, rolled over the thigh and are bound on a bamboo natwa. Traditionally, tasar cocoons are reeled by women. Hence, woman can reel 60-70 cocoons per day, which will produce about 50-60 gm of reeled silk.

The above method is replaced with improved technique of cooking of cocoons with the help of soap and hydrogen peroxide and introducing a power operated reeling machine consisting of four ends. The machine can insert upto 8-9 twist/inch (tpi) with the yield of 150-200 gm of reeled silk/8 hr. After steam setting the yarn produced by the machine can be used for weaving on handloom and powerloom with any further processing. Qualitatively such yarn can increase in weaving efficiency and fabric quality and compitable with Chinese tasar silk.

In India, Non-mulberry sericulture forms an important part of tribal economy, particularly in Eastern and North-eastern states of the country and Maharashtra. It also plays an important role in arresting the deforestation and generating gainful employment to the tribal population. India has abundant manpower and rich natural resources of food plants for wild silkworms. Therefore, there is need to utilise these two resources for the economical development of the country. The tribal population of India is over 40 million and nearly three-fourth is located in tasar producing states. Tasar sericulture is highly beneficial to tribal population since wild silkworm rearing is a low cost technology with practically no investment.

In Eri and Muga silk production India ranks first and second to China in Tasar silk. Both tropical and temperate tasar culture are practiced in India. Tropical tasar culture is concentrated mainly in tribal belts in the states of Bihar, West Bengal, Orissa, Madhya Pradesh, Andhra Pradesh, Maharashtra and Uttar Pradesh while, the temperate tasar culture is performed in the hilly regions of Uttar Pradesh, Himachal Pradesh, Jammu and Kashmir and North-eastern states. State wise, Manipur ranks first among the temperate tasar silk producing areas. Assam alone accounts for over 95 per cent of Muga and over 60 per cent of Eri silk. Average annual production of wild silk is estimated at about 2075 MT comprising 1530 MT of Eri, 428MT of Tasar and 117 MT of Muga silk during the years 2007-2008. In case of Muga and Tasar, annual fluctuations in production are common due to natural vagaries and as an outdoor activity. However, wild silk industry is facing to a number of problems and one of the major problem is lack of adequate market support.

Tribals harvest wild silkworms using forest trees, produce cocoons and sale in weekly haats at any prices offered. The producers are exploited by the commission agents due to monopoly. The traders (known as Mahajans) have links with buyers from other States. By the time, the cocoons get hoarded for long periods. There is another class of small time traders located around production areas who act as go between big traders and local weavers. Their modus operandi is like that of itinerant rural agents. Apart from operating in haat areas directly, these petty traders also buy cocoons from bigger traders cum financiers, supply to local reelers or weavers and retain good margin for themselves. The traders have no fixed practices or rules of pricing and hence, they can manage business.

However, during last two decades there has been tremendous improvement in the market system for wild silk products that Central Silk Board has provided alternative market support through establishment of two Raw Material Banks, one each for Tasar and Muga. The state governments also introducing regulatory and intervention measures to protect the interests of the tribals and to promote silk industry. Bihar, Orissa, Andhra Pradesh, Maharashtra etc, have established their own agencies for marketing of silk. While, other states like West Bengal and Uttar Pradesh promoted marketing environment without direct participation. This has, resulted in a competitive atmosphere to producers. Now each state has its own policies and priorities on wild silk industry.

In India Tasar silk products in various states are controlled by the Mahajans. There is some kind of traditional and financial link between trader-cum-financiers and producers under which producer have obligations to sale their cocoons only to the Mahajans or their agents whom they are indebted. Therefore, traders/Mahajans purchase cocoons at prices much lower than the prevailing market rates. This arrangement affects the interest of producers.

For dealing with wild silk products and cocoons every state in India has its own arrangement. Three centers have been established, by State Department in Bihar for tasar marketing which purchase tasar cocoons and dispose off the same to the reelers/ weavers and their societies. However, these centers are not fully equipped with facilities and resources to play effectively role in marketing.

In Orissa, tasar cocoons are not permitted to be either sold or purchased by any agency other than the Orissa State Tasar co-operative Society (co-operative federation) which holds monopoly, is unique feature of Orissa state in wild sericulture. The government sponsored co-operative societies procure cocoons through a network of procurement centers in Madhya Pradesh. In Andhra Pradesh, the

SERIFED, a federation sponsored by the state government operates through the Department of Sericulture for purchase and disposal of cocoons and yarn. The Development Corporation of Vidharba Ltd. (a state owned multipurpose development body corporate) provides market support to the producers in Maharashtra. While, in West Bengal, the Department of Sericulture procures only seed cocoons and supervises sale/purchase of commercial cocoons which is carried out at its selected centers. In Orissa and some other states, Govt. neither exercise monopoly nor restrict transaction of cocoons and allow market forces to operate. The governments only fix minimum floor price for cocoons.

With a view to provide alternative and remunerative market and also to regulate the prices for cocoons, the Central Silk Board set up a Raw Material Bank (RMB) for Tasar in Chaibasa (Singhum District of Bihar) in 1972. The RMB is operated through its main office at Chaibasa, supported by two sub-depots one each in Raigarh (Madhya Pradesh) and Bhagalpur (Bihar). Initially, the RMB and its units did provide a formidable competition to traders, but over the years, there has been a gradual reduction in their direct market purchases.

Presently, its overall share in the market is not significant. However, the RMB is not in a position to meet the full requirement due to meager supply. The RMB has facilities for long term storage, reeling and spinning as a standby arrangement. The TRIFED (A Government of India sponsored agency) is also engaged in purchase and processing of cocoons/yarn and fabric production. It procures cocoons from RMB and co- operatives operating in respective states. Its annual cocoon and yarn transaction is too small to influence the market. The Khadi and Village Industries Commission and Khadi and Village Industries Board in respective states have promoted co-operative societies which also buy cocoons from producers to some extent.

For ensuring remunerative returns to producers, Price Fixation Committees have been constituted in every State of India. Although, prices are fixed on different criteria, each state follows different norms. In Bihar, which accounts for highest production of tasar silk, the Price Fixation Committee fixes prices for different types of cocoons on the basis of the norms fixed by the RMB, Chaibasa.

In Orissa, cocoon prices are fixed on the basis of silk content and are quoted per tola. For fixing price, the Apex Society has evolved a formula under which weight of 5/10/20 samples (80 cocoons each) is taken. Prices are based on labour component of production and cost of cocoons in Madhya Pradesh. In West Bengal and Maharashtra, generally the prices are fixed on the basis of grades fixed by RMB, Chaibasa. Cocoons are classified into four grades depending on the quality in Andhra Pradesh. Central Silk Board (through RMB) fixes prices for different types of cocoons for procurement from producers or government agencies periodically.

Indian Wild Silk Export

Non-mulberry silks are abundantly found in remote regions, on hill tops and in forest interiors of Myanmar, China, India, Korea and parts of equatorial Africa and South-East Asia. Although entomologists visualized five hundred types of moths spinning silk cocoons, only a few have commercial value. Tasar, Eri and Muga are the three commercially known wild silks. Wild silks which accounts for 8.23 per cent of the total silk production is of considerable importance in the Indian context.

The versatility of Indian wild silk is unmatching. The silk fabric produced by our traditional craftsmen is found increasingly accepted by the changing fashionable world of Europe and USA. Fashion means variety, fantasy, innovation, vitality and versatility. All these are the hallmark of Indian wild silks. The vibrant colours and wide spectrum of woven designs fascinate the fashion conscious westerners which bear testimony to the skill, artistry and aesthetic sense of our weavers. Tasar silk has greater commercial value especially in exports and India is next only to China in Tasar silk production.

Status of Non-mulberry Sericulture in India

Non mulberry silks are cultivated in about 16 states of India (Figure 1). The status of non mulberry silk of important states of India is given below.

Andhra Pradesh

The host plants of tasar namely, *T. tomentosa* and *T. arjuna* are plenty in the forest areas of Warangal, Karimnagar and Adilabad districts and to some extent in Khammam and Mahaboobnagar districts. Therefore, Andhra Pradesh has good potential for tasar silk production. Tasar culture is mostly practiced by the tribals for the past few decades. Recently, the government of Andhra Pradesh provided several incentives to tasar rearers to improving tasar business. Tasar seed is being supplied to the rearers at a nominal cost of Rs.50 per 100 dfls. State Government provides forest areas having host plants are allotted to rearers free of cost for conducting tasar rearing. Market facilities are also provided by the department and reviewed seasonally by the committee constituted under the Chairmanship of the Director of Sericulture which decides the price of cocoons from time to time.

Andhra Pradesh has a remarkable trivoltine local tasar race. The level of production of tasar in the state (by end of 1997-98) was about 35.08 lakh no of cocoons. However, the third crop was badly affected by virosis and pebrine due to continuous rains and adverse climatic conditions during 1996-97 and the production was felled steeply to 35.08 lakh cocoon as against 61.35 lakh no of cocoons.

The requirement of tasar seed of Andhra Pradesh for first crop is about 1.25 lakh no. of dfls, while it was about 1.75 and 4.75 lakh no. of dfls for second and third crops, respectively. However, production of disease free quality tasar seed suitable to local conditions has become a burning problem. The quality seed for 3rd crop rearing is crucial as the returns depend on this crop and is wanted in Andhra Pradesh, Central Silk Board (CSB) is looking after this matter. However, some productive tasar races are maintained and quality seeds are supplied to rearers in the state through the departments such as Warangal, Karimnagar and Adilabad and Khammam districts. Liquidation of different varieties of tasar cocoon stocks procured from rearers is problem of Andhra Pradesh. A planned and comprehensive approach of tasar reeling cocoons is the need of the hour to sustain the tasar industry in the state. Considering the advantages of the industry, which help the poorest population in our society, tasar needs an articulate attention for its sustainable development on a large scale.

Bihar

Bihar state ranks first in the Tasar silk production and second in Eri silk production. However, on account of non-remunerative prices and non-availability of quality seed cocoons at fair price the acreage, productivity and production of silk has a declined (Sinha, 1998). The problems of weaving community are also unanswered besides, the presence of co-operatives.

In Bihar, the tasar growers are confined to remote villages of Chhotanagpur area comprising of Singhbhum, Ranchi, Gumla, Lohardaga, Palamu, Chatra, Giridih and Dhanbad districts and also in Santhal Parganas, Dumka, Deoghar, Sahebganj and Godda districts.

Zarkhand is also visualized as tasar culturing areas by districts like Banka, Jamui, Munger and Rohtas. About, 50,000 families are involved in tasar sericulture bussiness. The state sponsored institutions such as Eri Research Station Ranchi, Seed Multiplication Centre, Valmikinagar (West Champaran), Eri Seed Supply Stations, Barari (Bhagalpur), Mohamadpur Gokul (Muzaffarpur), Sirsa (Gopalganj), Ranchi, Bakhtiarpur (Patna) and Begusarai and 48 Resam Demonstration Centers are working for the development of eri silk. The above institutes produce and supply disease free layings to rearers and also train them in rearing and spinning of Eri pierced cocoons. As regard to eri culture about, 10,000 families are involved in this bussiness. During 1997-98, the production of Eri pierced cocoons was nearly 40 MT.

The Tasar and Eri yarn productions and silk provide self employment to the rural people. As such, more than one lakh families of the state are involved in various activities of wild sericulture *viz.*, cocoon production, reeling/spinning of silk yarn, weaving of pure and mixed fabric and marketing, etc.

Arunachal Pradesh

Due to suitable climatic condition available, land scape and growing interest of the people the wild sericulture in Arunachal Pradesh has an immense potentiality. The sericulture was introduced in the state as early as 1960 but, industry was unknown till recently due to lack of technical knowledge, infrastructure and institutional support. But, industry has remarkably developed with the implementation of various developmental programmes in an extensive way covering most of the districts of the state.

Eri culture is taken in the foothills of Papum Para, East Siang, Upper Siang, Lohit, Dibang Valley, Changlang and Tirap districts with increased production of Eri raw silk. Therefore, 17 Eri seed production cum demonstration centers are established in Arunachal Pradesh which supply disease free silkworm seed cuttings/saplings to growers and providing short term training to the beneficiaries and performing demonstration and motivational works. However, the development is not supposed to be so satisfactory due to lack of proper infrastructure, technical manpower and non-availability of Eri seed grainages. For procurement and supply of disease free seeds, the state government is dependant upon the neighboring states.

In Arunachal Pradesh muga rearing has been introduced in the areas of Nongkhon (Lohit district), Bordumsa (Changlang district) and Dibang valley. The villagers are gradually taking interest to take up muga rearing during agricultural off seasons. Oak tasar rearing has also been introduced in the villages of Dirang and Jerigoan areas of West Kameng district but, no satisfactory development take place due to non availability of quality seeds.

Maharashtra

In Maharashtra, tasar cultivation is traditionally practiced by Dhiwar communities in the districts of Bhandara, Chandrapur and Gadchiroli (Figure 2) since 200 years. The tasar development Corporation of Vidharba Limited, Nagpur has taken keen interest in tasar sericulture. The Government of Maharashtra implemented the Inter State Tasar Project with the assistance of Swiss Development Co-operation in above districts of Maharashtra. In Maharashtra, Sericulture was in hand of State Khadi and Village Industries Board, Development Corporation of Vidharba Limited, Nagpur and Directorate of Industries. However, for giving fillip to tasar culture a separate Directorate of Sericulture has been developed under administrative control of Textiles ministry of Govt. of Maharashtra.

1. In Maharashtra, host Plantation area under natural food plantation in forest and revenue sector is about 4000 ha.
2. The total area under economic food plantation under departments is 500, 400 and 200 ha in Chandrapur, Gadchiroli and Bhandara, respectively.
3. The economic plantations of *T. arjuna* raised by the forest department is 205, 313 and 263 ha in Chandrapur, Gadchiroli and Bhandara, respectively.
4. A modern cocoon market complex is established at Armori, for purchase of cocoons from the rearers and storage.

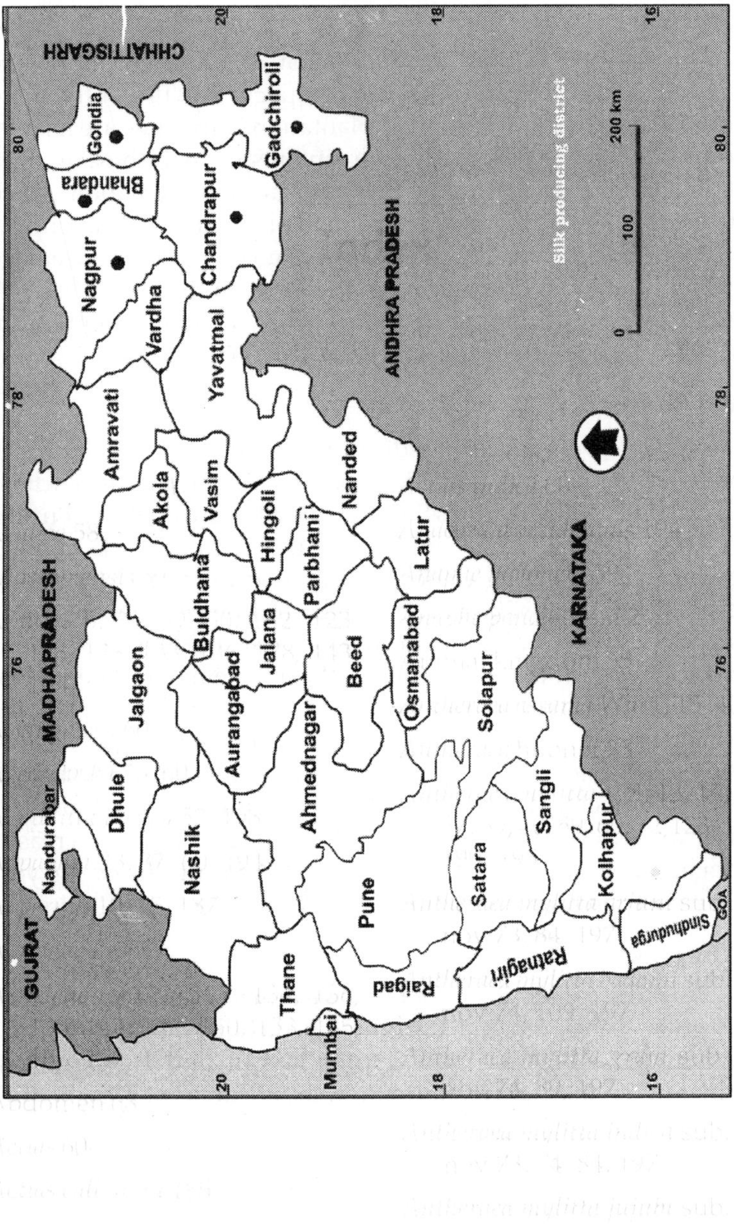

Figure 2: Tasar Silk Producing Districts of Maharashtra

5. State Govt. developed five tiny reeling units and one median reeling unit.

Production

During 1997- 98 the production of dfls and cocoons were recorded as 1, 30,186 and 44, 89,400 respectively. Presently, more than 4000 ha of natural tasar food plantation in the state are known. Around 2000 skilled tasar rearers are interested in the business but, not able to take up rearing by this natural food plantation due to Forest Act. Likewise there are 781 ha food plantation, available under Social forestry scheme. About, 550 beneficiaries are getting advantage through tasar silkworm rearing in the state. In the year 2006-2007 the raw silk production in Maharashtra was 85 mts mainly from tasar (50mts) and eri 35mts, (Jadhav *et al.*, 2007).

West Bengal

The state west Bengal of India produces all the three varieties of Non-mulberry silk *viz.* Tasar, Eri, Muga. The silks are mostly practised by tribal people. The districts Birbhum, Burdawan, Bankura, Purulia and Midnapur are famous for Tasar culture. About 7443 families are engaged in rearing under a cultivated area at 15,201, acres (Pertin,1998). Tasar silkworm is reared outdoors mainly on Arjuna trees besides Asan, Sal and Sidha (*Lagerstroemia*). Sukinda trivoltine races are being utilised for rearing in West Bengal. The schedule for 1st crop is from June 15 to 30 with a requirement of one lakh dfls, the normal period of 2nd and 3rd crop would be between August 18 to 31 and October 25 to November 8, with a requirement of 1.75 and 4 lakh dfls, respectively. The trend of dfls: cocoon ratio for 1 st, 2nd and 3rd crops are 1:15, 1:12/15 and 1:35/40, respectively. The increased estimated demand of the state is over 50 per cent dfls by the year 2000 have been found encouraging in West Bengal.

Reeling Sector

West Bengal has a strong tasar weaving base. Raghunathpur in Purulia, Tantipura and Karidhya in Birbhum, Bishnupur and Sonamukhi in Bankura and Gopiballavpur and Nayagram in Midnapur are the main reeling cum weaving areas of West Bengal. Recently, improved reeling machines have been adopted for increasing the silk activities. The State timely supplies required number of quality dfls particularly during 3rd crop. However, steady declination of leaf yield (both qualitatively and quantitatively) in Arjuna Block Plantation have been occurred due to lack of cultural operation and manuring as the farmers can not afford to provide necessary inputs due to their poor economic condition. In general, West Bengal is progressing the wild silk activities in the state.

Orissa

In Orissa, Tasar and Eri are produced as non-mulberry silks. Tasar culture is a tradition for the tribals of the Mayurbhanj, Sundargarh and Koenjher districts of Orissa. It provides an additional income to tribals in different seasons. The mention of tasar in Orissa can be found in epics, fables, folklores and historical recordings including palm leaf inscriptions. There are three groups of people–adivasis (tribals) collecting tasar cocoons, pataras (weavers) processing the same to fine fabrics and sadhabas who sailed overseas in their indigenous ships carrying these fabrics along with other items for trade. A systematic administrative management for tasar culture was first came in Mayurbhanj, Orissa. It received royal patronage from the Bhanja dynasty (640-1952 AD).

Tasar culture has been stimulated in the state largely during the second five-year plan. Co-operative Societies (TRCS) of tasar rearers were started during this period. At present there are 62 TRCS functioning in the state with the membership of 40,442 rearers. In 1962, an apex society, Orissa State Tasar and Silk Co-operative Society (OST&SCS) was established, has acquired monopoly over tasar marketing for longer period. Later, a good infrastructure has been built up with a common objective of the development of tasar culture and tasar rearers in the state. Seed stations, pilot project centers, grainage houses, reeling and spinning centers, weaving centers etc have also been encouraged at both government and co-operative levels. Five Basic Seed Multiplication and Training Centres, five Pilot Project Centres, four Tasar Seed Stations, three Compact Block Tasar Rearing Units and one Research Station, etc are working at present in Orissa.

Tasar in Orissa is represented by four major ecoraces–Modal and Nalia are wild ecoraces of *A. paphia* and Daba and Sukinda of *A. mylitta* are cultivated ecoraces maintained by Central Tasar Research and Training Institute, Ranchi. Bogal from Modal and Nalia seed cocoons have been cultivated commercially by tribals of

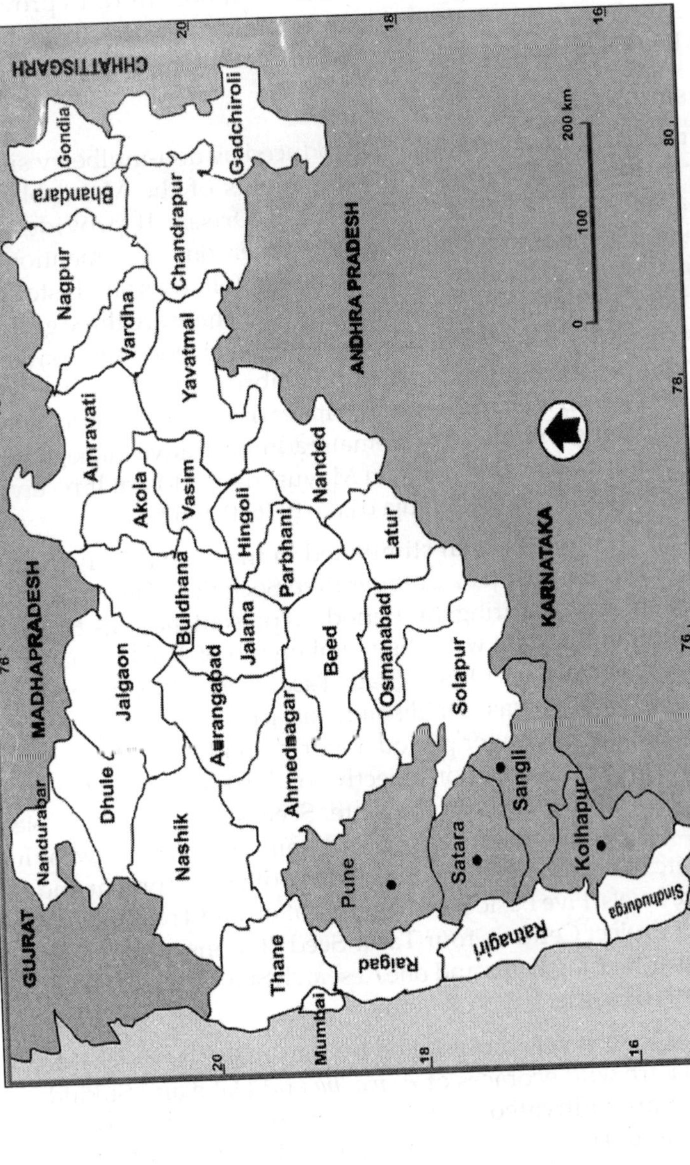

Figure 3: Map of Maharashtra Showing Study Area (Western Maharashtra Districts: Kolhapur, Sangli, Satara and Pune)

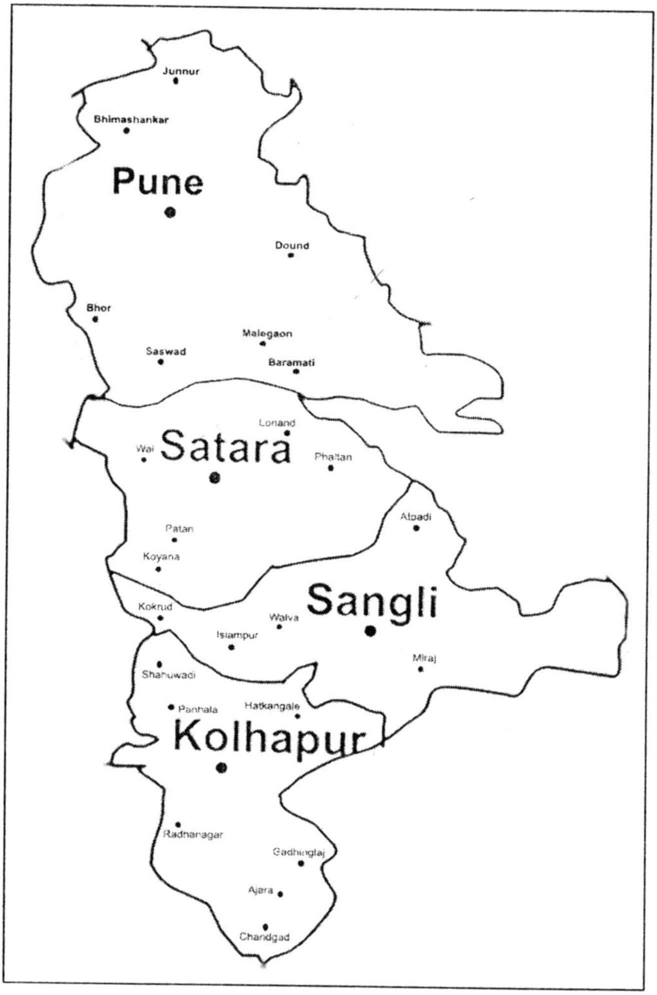

Figure 4: Map of Study Area (Districts: Kolhapur, Sangli, Satara and Pune Showing Collection Spots)

the state. It is the modal, the highest silk bearing cocoon in the world and is abundant in the forest from North to North-West Orissa. These ecoraces have different voltinism at different altitudes. The primary food plants include *S. robusta* (Sal), *T. tomentosa* (Asan) and *T. arjuna* (Arjun). In addition, a number of secondary food plants have also been recorded for above ecoraces being an important wild silk state of India. The activities related to wild silk should be strengthened in future. In addition to above several states of India representing wild silks and there is need to exploite the wild sericulture technique on large scale.

Considering the importance of wild sericulture in socio-economic development of tribal population, the present topic is selected for the betterment of the people.

Chapter 2

Review of Literature

Review of literature indicated that out of four varieties of silkworms viz, *B.mori, A.mylitta, A. assama* and *P.ricini* the literature is upto date on *B.mori,* quite sound on Muga and Eri but, so far as tasar (*A. mylitta*) is considered, the reports are haphazard, unsystematic, discontinuous and inconsistent because of its wild cultivation by the tribal people. More significantly, a systematic, scientific and molecular approach is given to the discipline of mulberry sericulture but, the literature on tasar is regrettably inadequate though some preliminary and baseline studies are carried out on semi- domesticated varieties *viz., A. mylitta* and *A. paphia* L.

Biodiversity of moths have been scarcely attended (Hubner, 1819; Westwood, 1848; Hampson, 1876; Moore, 1877) at Global Scenario. Cotes (1891-1893) studied the wild silk insects of India. Jordan (1911) described a new Saturniidae moth. Similarly, Watson (1911) studied the wild silkmoths of the world with special reference to the family Saturniidae. Potter (1941) reported the Chinease moon moth *A. selene*. Collins and Weast (1961) studied Saturniidae wild silkmoths of United States. Sen and Jolly (1971) studied the genitalia of tasar silkmoth, *A. mylitta*. Jolly (1976) reviewed taxonomic hybrids related to *A. proyeli*. Barlow (1982) provided an introduction to moths of south East Asia. Thangavelu (1992) opined that there is a need for conservation of wild sericigenous insects of India. Thangavelu (1992) enlighted recent studies in Indian tasar and other wild silkmoths. Species diversity in tasar silkmoths, hammock formation and mortality of tasar silkmoth along with some tribal belief have been reported by Mohanty *et al.* (1995, 1996, and 1997). Narasimhanna (1998) described biodiversity of wild silkworms in India.

Mohan Rao and Satpathy (2000) reported modal –A unique tasar ecorace. Kirsur and Krishna Rao (2003) studied wild silks of India and their brand identity while, Satpathy and Mohan Rao (2003) studied tasar ecorace –Bogai, *A. mylitta* with respect to cocoon characters and production technology. Prasad and Nagaraju (2003) reported a comparative phylogenitic analysis of full length mariner elements isolated from the Indian tasar silkmoth, *A.mylitta*. Shankar Rao *et al.* (2004) worked with ethology and conservation of raily ecoraces of Indian wild tasar insect, *A. mylitta*. Rout *et al.* (2004) advocated for the need of conservation of tasar ecorace modal. Recently, Sathe (2000, 2004) contributed on biodiversity of wild tasar silkmoths from western Maharashtra, India. He reported 13 species of tasar moths from western Maharashtra. Suryanarayana and Srivastava (2005) prepared a monograph on tropical tasar silkworm under which he reported 44 ecoraces.

India's wild silkmoth biodiversity consists of 47 species, 15 genera, 3 tribes and 2 sub–families of the family Saturniidae (Chaoba singh and Suryanarayanna, 2005). Sahu and Bindroo (2007) described wild silkmoth biodiversity in the North eastern region of India and stated the need for conservation of Tasar silkmoths.

Taxonomical details of wild silkmoths are very scanty in Indian subcontinent and oriental region. Taxonomical work of wild silkmoth refers to Linneous (1766), Leach (1815), Hubner (1818), Helfer (1854), Moore (1859, 1872, 1877, 1888, 1890), Hampson (1892), Westwood (1959), Walkar (1886), Seitz (1913) Tutt (1899-1909) FAO (1987), Sathe *et al.* (1997, 2002, 2004) etc. Hampson (1976) described 7 species from India. Nayak and Dash (1986) reported the sex association in double cocoons of *A.mylitta*. Nayak *et al.* (1987) for the first time classified the tasar cocoons on the basis of pupa, shell, peduncle and loop.

Sathe (2007) described 13 species of wild silkmoth from Western Maharashtra. No taxonomical work is reported in recent years on wild tasar moths except the work of Sathe (2007) and Kavane and Sathe (2009).

Rajadurai and Thangavelu (1998) studied the biology of moon moth *A.selene*. Saikia and Handique (1998) studied the biology of a wild silkmoth *A. atlas*. Similarly, Shankar Rao *et al*. (2002a) studied the biology of raily ecorace of *A. mylitta* a wild silkmoth while, Jayaprakash *et al*. (2002) studied the biology of Andhra local ecorace tasar silkmoth of *A. mylitta*.

Hamamura (1959) and Hamamura *et al*. (1962) reported nutritional requirement and ecology of the tasar silkworm *A.mylitta*. The proper selection of the host plants is of fundamental importance so far as consumption, digestion and assimilation of host plant leaves and production of cocoons are concerned. Sengupta *et al*. (1981) studied rearing performance of tasar silkmoth *A.mylitta* in crop III. Different aspects of cocoon such as cocoon colour, structure, cocoon formation in different environmental stress have been investigated in natural environment by Hayshia (1959), Figimato and Naomoso (1961),Narayanan and Sonwalkar (1967),Jayaswal (1973), Ghosh and Lamba (1976), Srivastava and Ansari (1979), Ghosh and Sengupta (1980), Devaiah *et al*. (1985), Sonwalkar and Jolly (1985), Sonwalkar (1986), Barah *et al*. (1988),etc. Findings on food plants and consumption of leaves in various larvae are recorded by some seriocologists, Nayak et al, (1985). Thangavelu *et al*. (1992) described indoor rearing of tropical tasar silkworm *A.mylitta*. Ojha *et al*. (1996) studied various aspects of tropical tasar silkmoths species. Alam *et al*. (1998) made the survey, collection and characterization of modal ecoraces of Indian wild silk insect, *A.mylitta*.

Saxena *et al*. (1997) studied the impact of a biotic factors and microclimates on cocoon and seed productivity of *A.mylitta*. Shamitha and Purushotham (2002) added on comparison between out door and total indoor rearing of an Andhra local ecorace of tasar silkworm, *A, mylitta*. Prasad *et al*. (2004) reported cashew as another food plant for tropical tasar silkworm. Mahobia *et al*. (2005) studied on different populations of raily ecorace in different ecopockets of baster forests in Chhattisgarh. Prasad *et al*. (2005) described indoor rearing of tasar silkworm. Singh and Bindroo (2006) suggested ideal rearing sites for Oak tasar culture in North-western Himalayas. Shimatha (2007) reported total indoor rearing of the tasar silkworm *A. mylitta*.

Studies have been made on the occurrence of voltinism and the effect of temperature and photoperiod of different *Antheraea* species by several workers. Noteworthy amount them are Dawson (1931) Tanka (1950a, b, c), Mansingh and Smallman (1967), Jolly *et al.* (1974), Kumar and Sinha (1974), Kato and Sakate (1981), Williams *et al.* (1985), Bhal *et al.* (1987), Jolly *et al.* (1987), etc. Environmental regulation of voltinism in *A.mylitta* has been studied by Nayak and Dash (1991).

Chakrovoty and Das (1979) studied the diseases of tasar food plants and their control measures. The sex association in bishellate cocoons of tasar silkworm has also been reported by Nayak *et al.* (1988). Tiwari (1997) opined that the arjun plantation under social forestry is highly effective for the larval growth of tasar. Recently, various aspects like biodiversity and sustainable development, egg laying behaviour and certified egg production, sustainable utilization and development through tasar biodiversity, bio-types of tasar cocoons, conservation of modal ecoraces including tasar culture and eco-friendly attributes have been accounted in detail (Mohanty and Dhal 1997, Mohanty and Mohanty 1997, Mohanty *et al.*, 1997 b, Mohanty and Behara 1997, Mohanty *et al.*, 1998, Mohanty and Behara 1998, Mohanty 2002).

Various abnormalities in cocoons, particularly double cocoons and its classification have been studied in *A. mylitta* and *A. assama* by Rao and Rao (1961), Kumarraj (1968), Saxena *et al.* (1969), Siddique (1989) and Naraian *et al.* (2002). Goel *et al.* (1993) and Mohanty *et al.* (1997a) reported sustained socio-economic development through tasar culture. Satpathy *et al.* (1985), Tiwari (1985),Sinha and Sinha (1994), Mohanty (1995), Mohanty *et al.* (1997a, 1998) and Mohanty (2000, 2001) have suggested the strategies for conservation of tasar fauna and tasar related flora in India. Kapila (1990) visualized important background and area of tasar culture through social forestry. Dash *et al.* (1992) studied oviposition behaviour of Goda modal ecoraces of Indian wild silkmoth *A. paphia*. Sahay and Kapila (1992) demonstrated the adverse effects of deforestation on tasar culture. Patro *et al.* (1997) suggested the usage of vegetable disinfectants for tasar eggs. The migratory behaviour of *A. paphia* in nature was observed by Patro and Dash (1998) while, Chaudhury and Sinha (1994) studied the moulting behaviour of tropical tasar silkworm *A.mylitta*.

Danilevskii (1965) proposed the aspects of photoperiodism and seasonal development of silk insect. The details of abnormal cocoons and sex ratio of colour cocoons in different seasons of tasar silk insect *A. paphia* have been discussed by Mohanty (1994a, 1994b). Akai (1998) contributed on global scenario of wild silk. Das (1998) studied colouration in wild silkworms. Sinha (1998) reviewed the research activities on wild silk in India and ecorace of tasar. Shetty and Samson (1998) reported status of non-mulberry sericulture in India. Sinha (1998) contributed non mulberry host plant maintenance. Nagaraju *et al.* (1998) studied biotechnological perspectives in non-mulberry silkworms. Tiwari (1998) discussed the advantages of typical tiwari grainage tray for better seed production. Patil (1998) studied various aspects of *A.mylitta* tasar silkmoth. Ghosh (1998) reported improved technique for wild silk cocoon processing while, Krishna Rao *et al.* (1998), Koshy (1998), Jahagirdar (1998) etc exposed socioeconomic and marketing aspects of wild silk.

Prasad *et al.* (2002) Studied characterization and molecular phylogenitic analysis of mariner elements from wild and domesticated species of silkmoths. Shankar Rao *et al.* (2002b) visualized a modal for nature grown tasar conservation. Shankar Rao *et al.* (2002c) demonstrated current technology and tasar cocoon productivity in baster. Verma *et al.* (2003) contributed on contribution of adopted seed rearers in tasar seed production. Mohan Rao *et al.* (2003) developed a technique for a modal, a unique tasar ecorace. Ahmed (2003) explained the role of participation of women groups in development of tasar culture. Singh *et al.* (2004) studied *Terminalia* wood ash as in alternative disinfectant in tasar culture. Mohan Rao *et al.* (2004) added on grainage and rearing behaviour of modal in ex situ conservation. However, Sonwalkar (2004) suggested that there is need for international standards for Indian raw silk.

Satpathy (2004) proposed integrated package of input requirement for tasar culture. Kimothi (2004) reported Oak tasar in Uttarakhand–towards improving productivity norms. Shankar Rao *et al.* (2005) studied reproductive potential of raily tasar silkworm was under natural and captive conditions. Sinha *et al.* (2007) demonstrated the effect of fortification of leaves of *T. tomentosa* with secondary nutrients of tasar silkworm *A. mylitta*. Ghosh *et al.* (2007) demonstrated on application of lac dye on tasar silk. Mohan Rao *et al.* (2008) studied forest protection force and tasar development. Gupta *et al.* (2008) reported that the tasar culture is the way of economic development. Kumar *et al.* (2009) explained the status of tasar flora and their potential utilization in India.

Sengupta (1987) reviewed the current status, scope and importance of tasar culture in India. Thangaveulu (1991), Sinha *et al.* (1992), Mohanty (1994c), Mohanty and Mohanty (1997), Nayak and Dash (1997) exposed problems and prospects of tasar culture in India. Mathur *et al.* (1998) suggested integrated package of rearing and the transfer of the technology to the field. Saxena *et al.* (1997) studied the impact of a biotic factors and microclimates on cocoon and seed productivity of *A.mylitta*. Shree *et al.* (2000) examined the post infectional biochemical changes in the leaves of tasar food plants. Pattanayak *et al.* (2000) studied allocation of energy for cocoon and pupal life of matured larva of *A. mylitta* the Indian tasar silk insect.

Rath *et al.* (2001) studied the effect of pebrine infection of fecundity and egg retention in silkmoth, *A. mylitta* in different seasons. Singh *et al.* (2001a) developed new technology for oviposition in oak tasar and its economics. Singh *et al.* (2001b) studied Oak tasar culture in Uttaranchal state, they reported role of appropriate technology of tasar sericulture in India. Shankar Rao (2001) studied tasar crop planning in Baster plateau of Chhattisgarh. Narain *et al.* (2002) exposed a new concept of seed organization in tasar.

Kishore *et al.* (2002) studied tasar food plants and silkworms attacked by large number of pest, parasites and predators. Chakrovorty *et al.* (2004) studied effect of out door preservation of tasar food cocoon on pupal survival and reproductive behaviour of Daba ecoraces of *A. mylitta*. Shankar Rao *et al.* (2004) described ethology and conservation of raily ecoraces of Indian wild tasar insect, *A. mylitta*.

Chandra and Gupta (2005) explained tasar culture and its role for human resource development. Rajany *et al.* (2005) discussed prospects of tasar culture in Kerala. Rai *et al.* (2006) described tasar sericulture as an emerging discipline for conservation and sustainable utilization of natural resources. Kakati (2006) discussed problem and prospects of Vanya sericulture in Nagaland. Kalantri *et al.* (2007) discussed progress and prospects of sericulture industry in Maharashtra. Dikshit (2007) visualized Similipal biosphere reserve a natural habitat of modal ecorace of *A.mylitta*. Singhvi *et al.* (2008) reported strategy for increasing tasar silk production in Maharashtra.

Chapter 3

Materials and Methods for Silk Technology

Technology counts the status of a country or a region and Sericulture is technology oriented business and prestigious biotechnology of common man in India. Application of wrong rearing technique results 100 per cent loss in the crop. Hence, technological advancements and modifications / ulterations are needed in improving the quality of product. Some times very minor change in technique can have drastic results, both positive and negative. For harvesting cocoons for commercial purpose, continuous rearing of silkworms is necessary. Therefore, in the present study, for rearing wild silkworms, following materials and methods have been adopted.

Rearing Materials

Rearing Boxes (Figures 5 a,b)

Plastic boxes were used for rearing early silkworm instars 1st and 2nd. These are translucent with detachable lid of size 27 cm × 6 cm × 7 cm (Figure 5a), and 39 cm × 30 cm × 21 cm (Figure 5b) in length, width and height respectively. The maximum capacity of box is 100 worms for rearing.

Rearing Trays (Figure 6)

The rearing tray was made up of metal material (galvanize)

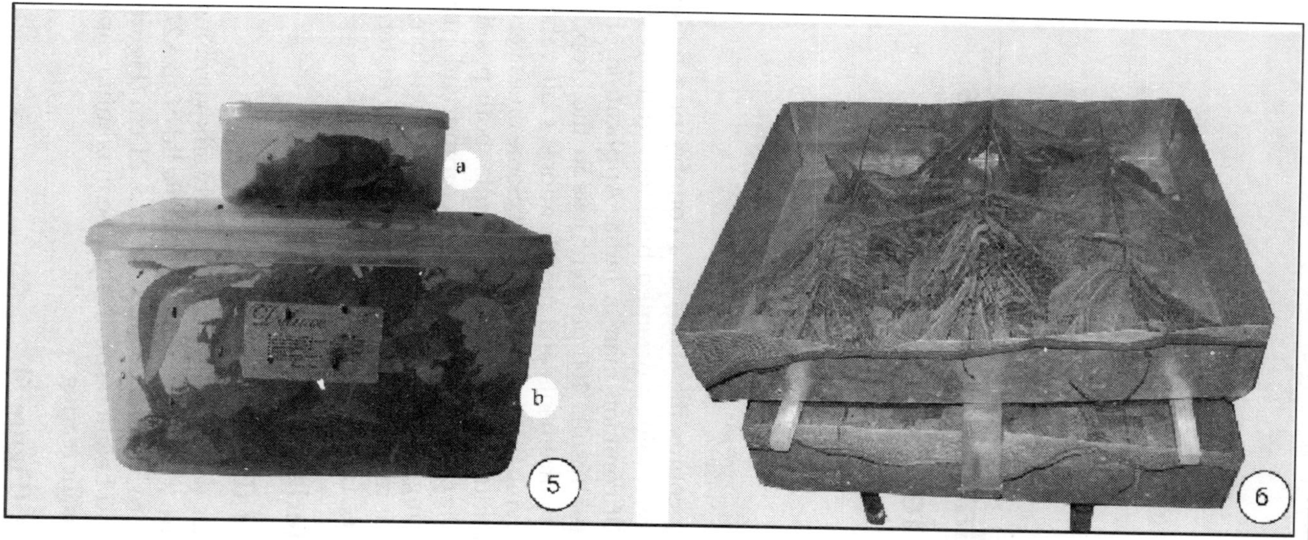

Plate 1

Figures 5a&b: Rearing Boxes; Figure 6: Rearing Trays

with size 3′ long × 2.5′ width × 6 cm in depth. Such type of trays were used for 3rd, 4th, 5th instars silkworms. The maximum capacity of each tray was 100 worms for rearing.

Rearing Room

Properly cleaned and disinfected room of the size 5 m × 4.5 m × 10' (length × width × height) with a single door and window was used for rearing of silkworms.

Wooden Cage (Figure 7)

Cocoons were kept in garland from hanging inside to the top of a wooden cage of the size 42 cm × 42 cm × 90 cm in length, width, and height respectively. There is fixed wire mesh on all the four sides including the door. It was used for protection to cocoons from enemies.

Egg Laying Box (Figure 8)

Egg laying box of size 12 cm × 6 cm × 4 cm in length, width, and height respectively was used for egg laying by individual fertilized female moth.

Incubator (Figure 9)

B.O.D incubator was used for incubation of eggs.

Oven (Figure 10)

Oven of the size 3.6′ × 2.4′ ft (height and width) was used for preserving and drying the moths.

Insect Cabinet (Figure 11)

Insect cabinet of size 7.5′ × 4.5′ × 3.4′ in length, width and height respectively containing 24 boxes, Vaman D. Purohit and Sons made was used for preserving the pinned moths. Napthalion balls were provided in each box at one corner for avoiding fungal and micro insect infections.

Insect Net (Figure 12)

For the collection of moths from the field, insect net was used. It is provided with aluminum handle of nearly 15 inch in length and circular metal ring of 6 inch in diameter and a collecting bag of 30 inch in depth made up of ordinary mosquito netting cloth attached to the iron ring.

Insect Boxes (Figure 13)

The standard insect boxes Vaman D. Purohit and Sons made of size 44 cm × 29 cm × 7.5 cm in length, width and height respectively were used for preserving moths. The insect boxes consists basal wooden portion covered with soft wood for pinning the insects and the upper side is perforated by glass and rest of the sides made up of wooden material.

Plastic Containers (Figure 14)

Plastic containers of size 16 cm × 6 cm in diameter and height respectively were used for killing the moths.

Insect Spreading Board (Figure 15)

Insect spreading boards (Figure 15) were used for pinning the moths.

Camera

Photographs of insects in the field were taken with the help of Canon power shot SX110IS camera (Made in Japan) having 9.0 mega pixel and 10 optical zoom. For close up photographs of moths a close up lens kit was used.

Hygrometer (Figure 16)

Hygrometer 'Paico' made was used for measuring temperature and humidity in the rearing room.

Nylon net

Nylon nets (mesh size 20 mm^2) were used to protect larvae from predators. The nets were applied over the tray.

Leaves of Silkworm Host Plants (Figures 17–21)

Leaves of Arjun (*T. arjuna*), Ain (*T.tomentosa*), Badam (*T. catappa*), Ber (*Z. jujuba*), and Angier (*Ficus carica* L) were utilized as diet for larvae. Freshly plucked leaves/ twigs containing a few leaves were washed with tap water and blotted before use as diet for the larvae. Only two feedings were given in 24 hr.

Chemicals

Following chemicals were used for preparation of slides and preserving insects.

Plate 2

Figure 7: Wooden Cage; Figure 8: Egg Laying Box

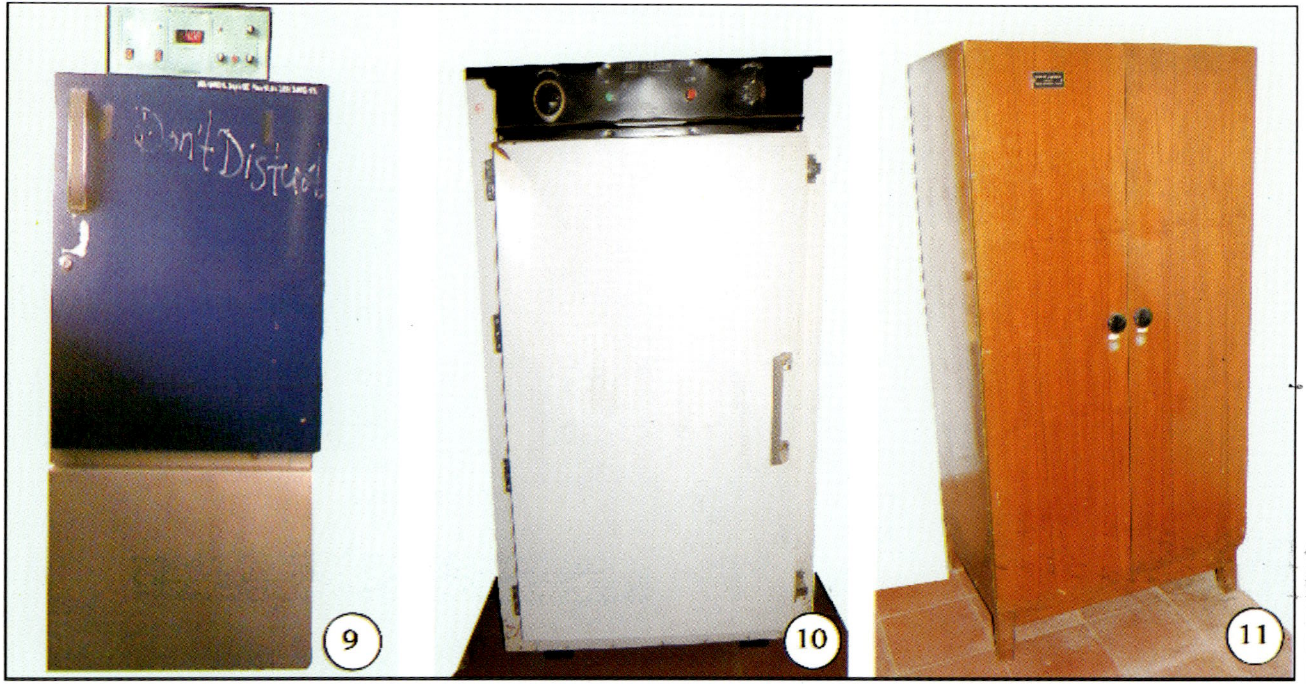

Plate 3

Figure 9: Incubator; Figure 10: Oven; Figure 11: Insect Cabinet

Plate 4

Figure 12: Insect Net; Figure 13: Insect Boxe; Figure 14: Plastic Containers with Moths; Figure 15: Spreading Board with Moths; Figure 16: Hygrometer

Plate 5

Figure 17: ***T. arjuna*****; Figure 18:** ***T. tomentosa*****; Figure 19:** ***T. catappa*****; Figure 20:** ***Z. jujuba*****; Figure 21:** ***F. carica***

1. 10 per cent KOH
2. 30 per cent to 100 per cent Ethyl alcohol grades
3. Glacial acetic acid
4. Xylene
5. DPX/ Canada Balsum.

METHODS

Moths were narcotized in ether/chloroform and killed in killing bottle. The moths were pinned with entomological pins from the dorsal of mesothorax and dried in oven at 55°C, the dried specimens then kept in wooden insect box. Few naphthol balls were placed at the bottom of the box for preventing fungal attack. Holotype and paratypes were pinned, labeled and kept in insect box. The records were made on locality, date of collection and identification. For the taxonomical studies antenna, genitalia were mounted on slides in D.P.X.

Preparation of Genitalia

By cutting and opening abdominal extremity the genitalia was removed. The dissected genitalia was kept overnight in 10 per cent cold potassium hydroxide solution to clear the muscles. (Keeping for a longer period of time in the Potassium Hydroxide (KOH) solution results in desclerotisation of the genitalic parts making it extremely difficult for study). Muscles were cleared with the help of needles, the cleared genitalia was then washed throughly in distilled water and preserved in 90 per cent alcohol in a microvial. This was either kept along with the specimen (if the specimens were preserved in sprit) or in case of a pinned specimen, it was kept separately and an index-numbered label was given to both the genitalia and specimen from which it was dissected out. Illustrations of the genitalia were made with the help of camera lucida. The procedure of affixing the genitalia to a piece of paper kept with the pinned specimen was not adopted since this makes it impossible to study the ventral view of the genitalia (Sen and Jolly, 1971).

Morphological studies were carried out with the help of monocular microscope. Comparative measurements of body parts of specimens were taken with the help of ocular micrometer and calculated with the help of graduated mechanical stage. All measurements were made in millimeter. The terminology adopted in the descriptions is same as described by (Hampson, 1894; Sen and Jolly, 1971; FAO, 1987; Sathe and Pandharbale, 2008) details of which is given in chapter 4, Taxonomy of wild silkmoths. A large number of references were consulted for the present work and cited in bibliography.

Photography

The whole wild silkmoths and their body parts such as genitalia were considered for photography. Whole insects and body parts such as genitalia were made with the help of Canon power shot SX110 IS camera (9.0 megapixal, 10x optical zoom). Whole mount figures have been enlarged 1x times to its original size, while other body parts have been enlarged to 115x times to its original size.

Rearing of *A. mylitta* Drury

Rearing of 1st and 2nd Instar Larvae

Newly hatched larvae were released on the leaves of the host plants (*T. arjuna, T. tomentosa* and *T. catappa*). With the help of soft camel hairbrush. 4–5 such leaves with mounted larvae were placed in the plastic box size, 27 cm × 6 cm × 7 cm in length, width, and height respectively (Figure 5a). The maximum portion of edge of each leaf was available to the larvae for feeding, the box was perforated with numerous exits for airation and with covered lid to prevent escape of the larvae. Next day, the larvae were transferred to new box containing fresh leaf diet. The moulting larvae transferred along with their support leaves. The old leaves were removed from the boxes at 12 hr interval. The used boxes were then washed, disinfected and dried for re-use.

Rearing of 3rd, 4th and 5th Instar Larvae

Rearing trays of size (3′x 2.5′x 6′in length × width × height respectively) were used for rearing of 3rd, 4th and 5th instar larvae. A twig having 12 to 15 leaves was used as leaf diet. 8 – 9 leaf twigs were introduced in the tray at a time. Larvae were transferred to a new tray along with help of new food. Touching with hand to food plants and silkworms was avoided mostly. The trays were cleaned after 24 hr interval. Moulting larvae were transferred along with the left over parts of the food plants.

Spinning of Cocoons

The full grown 5th instar larvae wandering for cocoon construction were sorted out and transferred to a new tray fitted with wooden rods and some old twigs that provided them opportunity to form the cocoons. The larvae used the space between the twig and the bottom of the tray for cocoon construction.

Collection and Preservation of Moths

Moths have been collected from different study spots of western Maharashtra with the help of insect net. The collected specimen were anesthetized for short period for their description. After making description, the species were released in the environment from where they were collected. The moths have been collected from Kolhapur, Sangli, Satara and Pune districts (Figure 4). Only a few species were killed by using chloroform in insect killing bottle and further these species were pinned by using spreading board (Figure 15) and dried in oven (Figure 10) and preserved in insect boxes (Figure 13). The terminology adopted in the description of species is same as described by Hampson (1894), Sen and Jolly (1971), FAO- Manual on sericulture (1987) and Sathe and Pandarbale (2008).

Survey and Surveillance

Survey of moths have been made at eight days interval in Western Maharashtra, *i.e.* Kolhapur, Sangli, Satara and Pune districts (Figure 4). The species have been noted/ collected at various spots in study area through out the course of study by one man one hour search.

Host Record

Host record of moths have been prepared by observing caterpillars feeding on particular plants in the western Maharashtra Kolhapur, Sangli, Satara and Pune districts. Later, the moth species and the plants have been identified by consulting appropriate literature.

Distribution of Moths

Distributional records of moths have been made by spot observation of the species by one man one hour search method.

Collection and Preservation of Larvae of Moths

Silkworm larvae of various wild silkmoths have been collected from different food plants from Western Maharashtra (Kolhapur, Sangli, Satara, Pune districts) and reared in the rearing room on the natural food material for adult emergence. The adults, thus formed have been considered for the description of the species.

Taxonomy of Moths

Taxonomical studies have been made on adult moth by consulting appropriate literature (Hampson, 1994a; FAO- 1987; Sen and Jolly, 1971; Srivastava *et al.*, 2003; Sathe and Pandarbale, 2008 etc). Taxonomical descriptions have been made with the help of compound microscope/ binocular/Lens/naked eye. Measurements of species body parts were taken in cm/mm in description.

Seasonal Abundance

Seasonal abundance of moths has been studied by spot observation of the species in different study centers at one week interval by one man one hour search through out the course of study. All seasons of the year have been considered for the study (summer, winter and autumn).

Chapter 4

Taxonomy of Wild Silkmoths

Taxonomy plays a very significant role in the field of applied biology including public health, national defense, pest management, forestry, environment problems, wild life management, nutritional science, forensic science and several other fields in identifying the species. Identification of species is first step in any experimental work. Hence, taxonomists are must. The subject taxonomy was treated unrewardly and stepmotherely by the financial agencies for long time in India. Secondly, molecular biology and biotechnology have attracted the attention of students. Hence, taxonomy has never been an attractive profession for the students but, now the picture is changed. Funding agencies are providing lot of funds for doing research on taxonomy because, taxonomists need in Research Institutes, Museums, Central and Government agencies, Industries, Zoos, Universities, etc. In fact, taxonomists are largely involved in designing and implementing control programmes of pest and diseases effectively than other subject experts. Taxonomy and biological and biotechnological sciences are interlinked. It helps in understanding evolutionary approaches of the species and guide the explorations for native and exotic species for fruitful use.

In the entire animal kingdom insects forms the largest group. They are divided into 40 Orders by Martynov (1938) and 30 Orders by Imms (1940) that are in turn broadly grouped into wingless (silverfish) and winged insects (bugs, butterflies, moths, beetles, flies etc) (Sathe,2005). Insects are the only invertebrates which can fly and seem to appearanced around 360 million years ago, around much before the dianosaurs appears on the earth. Insects 'boom' in the lush, gloomy and fern forests of the world. Insects are of the most successful groups of animal. They have significantly managed to survive the onslaught of Homo sapiens. Following the evolution of flowering plants, moths and butterflies (Order: Lepidoptera) were among the last to arrive on the evolutionary scene around 160 million years ago. According to Kehimkar (1997) there are about 25000 known species of butterflies and over, 1,20,000 moths. The combine number of moths and butterflies as Order: Lepidoptera is second to largest Order: Coleoptera (beetles and weevils) of class insecta. Much of the evolutionary history of these insects is guess-work, however, very few fossil records that have been found shown that the structure and pattern of the wing veins of the moths existed around 30 million years ago and appear very similar to those found today. Some primitive moths existing even today are believed to have a common ancestor with caddisflies. Such types of little moths have no proboscis for sucking liquid food but, instead have biting jaw to feed on pollen. Very strikingly indicating the early moths had biting mouth parts, some moths like Atlas and Tasar have no mouth at all as they do not feed in their very short life span of two weeks as adults (Kehimkar, 1997).

Moths belong to Order: Lepidoptera. They differ from butterflies by absence of clubbed antennae and have dull colouration than butterflies, and mostly they fly by night. Very typically, they keep their wings parallel on the sitting substrate. If mouth parts present in moth, are modified into long coiled tongue called as proboscis. The moth shows distinct stages of its life cycle viz, egg, larva, pupa and adult. The eggs of moths vary greatly in shape, size and colour. Tasar and Moon moths lay them in small clusters on the leaves of favourite food plants.

Moths are a distinct taxonomical group of the Order: Lepidoptera.

Moths in general shows following characteristics:

1. Medium to large sized

2. Covered with overlapping flat scales forming colour patterns.
3. Mouth parts modified into a coiled sucking proboscis.
4. 2 pairs of wings present
5. No cerci present.

According to Mani (1993) the Order Lepidoptera is divided into three suborders namely,

1. Jugatae
2. Rhopalocera, and
3. Frenatae

Jugatae: In Jugate wings are interlocked by jugum and venation of fore and hind wings are similar.

In Rhopalocera antennae are knobbed at tip or thickened before tip, wings are without frenulum but, with strongly arched and butterflies are included under this suborder.

In Frenatae antennae are simple or variable, rarely swollen at the tip and wings with frenulum.

The suborder Jugatae is divided into following super families, Micropterygoidea and Hepialoidea.

The sub Order: Frenatae contains following super families namely,

- ☆ Cossocideia
- ☆ Castnioidea
- ☆ Zygaenoidea
- ☆ Incurvarioidea
- ☆ Nepticuloidea
- ☆ Zygaenoidea
- ☆ Pyralidoidea
- ☆ Sphingoidea
- ☆ Elachistoidea
- ☆ Gelechoidea

- ☆ Yponomeutoidea
- ☆ Tortricoidea
- ☆ Pterophoroidea
- ☆ Tineoidea
- ☆ Uraniodea
- ☆ Geometroidea
- ☆ Drepanoidea
- ☆ Noctuoidea
- ☆ Saturnioidea
- ☆ Bombycoidea

and Rhopalocera contain two superfamiiies namely Hesperioidea and Papilionoidea. The super family Saturnioidea has a great economic importance since the individuals produce silk. The super family Saturnioidea contain only one family namely, Saturniidae. It is characterized by:

1. Antennae bipectinate in both the sexes, the rami being longer in males.
2. A prominent eye spot near the centre of each wing is present.
3. Hindwings with veins $Sc+R_1$diverging from the cell base M_2 (median vein) arising at or infront of middle of cell, frenulum completely absent.
4. The maxillae are absent and labial palps are small.
5. Larvae are stout and possess scoli of unequal size on all the segments.
6. Larvae are polyphagous and spin large and dense cocoons.
7. Some of the species belonging to this family over winter and complete one generation in a year whereas others are multivoltine in nature.
8. The members of this family are mostly found in tropical countries but, a few are present in temperate regions.
9. This family includes largest moths of the world having wing expanse of about 25 cm and are mostly conspicuous or brightly coloured, *e.g.*, *A. alias*, *A. edwardsi* etc.

10. It includes non-mulberry silk producing insects such as *Antheraea mylitta, A. proylei, A. roylei, A. yamamai, A. assamensis, S.c. ricini, Anaphe moloneyi, A. atlas, Pachypasa otus* etc. Besides, these, it also includes some non-commercial sericigenous lepidopterans like *A. selene* (moon moth) and *C. trifenestrata* (cashew caterpillar). The family Saturniidae contains two subfamilies viz, Salassinae and Saturniinae.

The subfamilies Saturniinae shows following characters.

1. Moths medium to very large sized with adult spans ranging from 7.5 to 15 cm.
2. The sub family Saturniinae is further sub divided into five tribes namely, Attacini, Saturniini, Bunaeini, Urotini and Miragonini. Out of which the two tribes *viz.*, Attacini and Saturniini plays very important role in producing good quality cocoons. The tribe Saturniini further sub divided into following important eight genera namely,

   - ☆ *Rhodinia* Staudinger
   - ☆ *Actias* Leach
   - ☆ *Saturnia* Schrank
   - ☆ *Loepa* Moore
   - ☆ *Cricula* Walker
   - ☆ *Lemaireia* Nassig and Holloway
   - ☆ *Solus* Watson and
   - ☆ *Antheraea* Hubner.

Taxonomic account on Saturniidae of Indian moths indicates that different workers such as Linnaeus (1758), Fabricus (1793), Guene's (1952) have named many species of Lepidoptera. However, Hampson (1976) made very significant contribution on Indian moths in Fauna of British India, moths – I, under which he described 23 families including the family Saturniidae. Under the family Saturniidae he described 3 species of the genus *Actias*, 7 species of the genus *Attacus*, 7 species of the *Antheraea*, 10 species of the genus *Saturnia*, 3 species of the genus *Loepa*, 2 species of the genus *Salassa* and 2 species of the genus *Cricula*. All above Saturniidae species have been reported from India. While according to Brown *et al.* (2007) under genus *Attacus* single species, under genus *Rhodinia* two species, under genus *Actias* seven species, under genus *Saturnia* seven species, under genus *Loepa* six species, under genus *Cricula* three species, under genus *Lemaireia* two species, under genus *Solus* one species and under genus *Antheraea* five species have been described from New World Tropics and Mexico. There are about approximately one dozen described species living in Europe, one of which, the Emperor moth, occur in the British Isles and 68 described species living in North America,42 of which reside north of Mexico and Southern California.

Recently, Sathe (2007) reported 2 species of the genus *Attacus i.e. A. atlas* and *A. edwardsi* from the genus *Actias* he reported two species namely *Actias indica, Actias paphia* out of which *Actias indica* is new species reported by the author. Under the genus *Antheraea* Sathe (2007) reported 5 species namely *A. mylitta, A. cingalesa, A.knyvetti, A.andamana* and *A.helferi.* He also reported *Saturnia anna, Laepa katinka* and *C. trifenstera* from Western Maharashtra.

Hampson (1894-1919) made a significant contribution to the noctuid fauna of India. In fact, the compiling of the taxonomic account of Indian moth species (besides Myanmar, Bhutan and Sri Lanka) in I to III volumes of "Fauna of British India" is an outstanding contribution made by the author.

In past, Hampson (1976), Arora and Gupta (1979), Sathe and Pandarbale (1999, 2004, 2008), etc, studied taxonomy of moths from India. Other taxonomical workers related to genitalia of tasar moths refer to Zander (1903), Pierce (1909- 1943), Snodgrass (1935), Viette (1948), Sen and Jolly (1967,1971), etc, while, ecoraces of *A.mylitta* have been studied by Thangavelu (1992), Alam *et al.* (1993), Narasimhanna (1998),Akai (1998), Satpathy and Rao(2003), Mohanty (2003),Mohan Rao *et al.* (2004), Kirsur and Krishna Rao (2003), Shankar Rao *et al.* (2004), Rout et al.,(2003), Mitra &Moon (2009) etc. Up to date 7 species of the genus *Antheraea* have been reported (Hampson, 1976; Sathe and Pandarbale, 2008) and 44 ecoraces have been recorded from India (Sinha, 1998; Srivastava *et al.*, 2003). However, no subspecies has been reported from the world. The present work will add great relevance in identification and phylogeny of the species.

Materials and Methods

For taxonomical studies silkmoths have been collected from the fields of Western Maharashtra at 15 days interval from May to January (2005 – 2009). The moths collected have been preserved in the insect box after pinning and drying. Taxonomical observations have been made on head, thorax, abdomen and their appendages. Measurements were taken with the help of ocular meter. Cocoons collected / reared have been characterized with respect to weight, shell weight, shell ratio, length, width, peduncle length, ring diameter, filament length and filament shade.

Preparation of Genitalia

For preparation of genitalia, the hind portion of the abdomen of male moth beyond 6th segment was separated out, boiled in 5 per cent KOH solution for about 30 min and then kept over- night in the same solution. This resulted in the removal of muscles and partial bleaching of chitinuous parts. KOH was neutralized by acetic acid treatment for about 30 min. The material being very big, examination of the unstained specimens in thick Canada balsam on slides under dissecting microscope was helpful in studying the genitalia. In describing the genitalia the terminology adopted was the same that of Snodgrass (1935), Sen and Jolly (1971) and Hampson (1976).

Morphological Considerations

Morphological characteristics of a typical moth are represented in Figures 22 to 28.

Head (Figures 22, 23a, 23b)

The dorsal skeleton is divided by two transverse sutures into clypeus (2) epicranium and occiput. The epicranium provides laterally the sockets for the antennae. The clypeus is the largest plate of the three. It is more or less strongly convex especially medially. It bears at the anterior margin of the labrum. The labrum is in most instance raised to a large transverse cariniform tubercle fronted over the base of the tongue (Hampson, 1902, Bell and Scott, 1976). The epistome covers the base of tongue, when normal it has a thin medial lobe and a large process at each side. The lateral processes are designated as "pilifer". The normal pilifer (5) is a curved obtuse process concave and flattened on the inner side and the inner surface

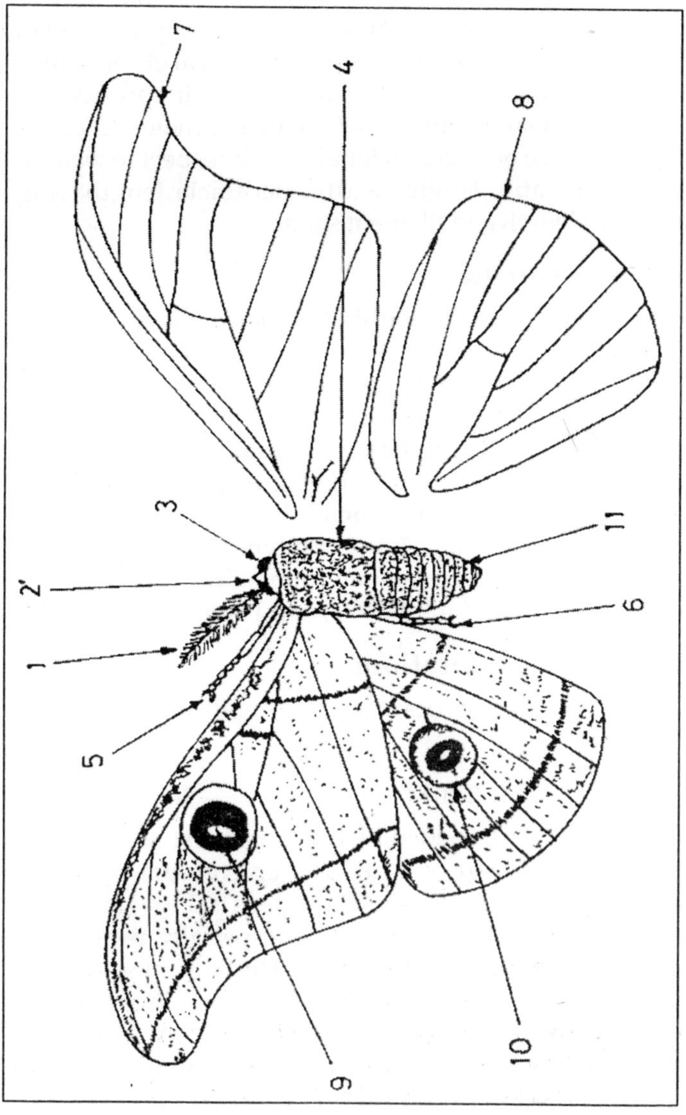

Figure 22: Adult Moth
1: Antenna; 2: Head; 3: Eye; 4: Thorax; 5: Fore leg; 6: Hind leg; 7: Fore wing; 8: Hind wing; 9: Hyaline area; 10: Ocellus; 11: Abdomen

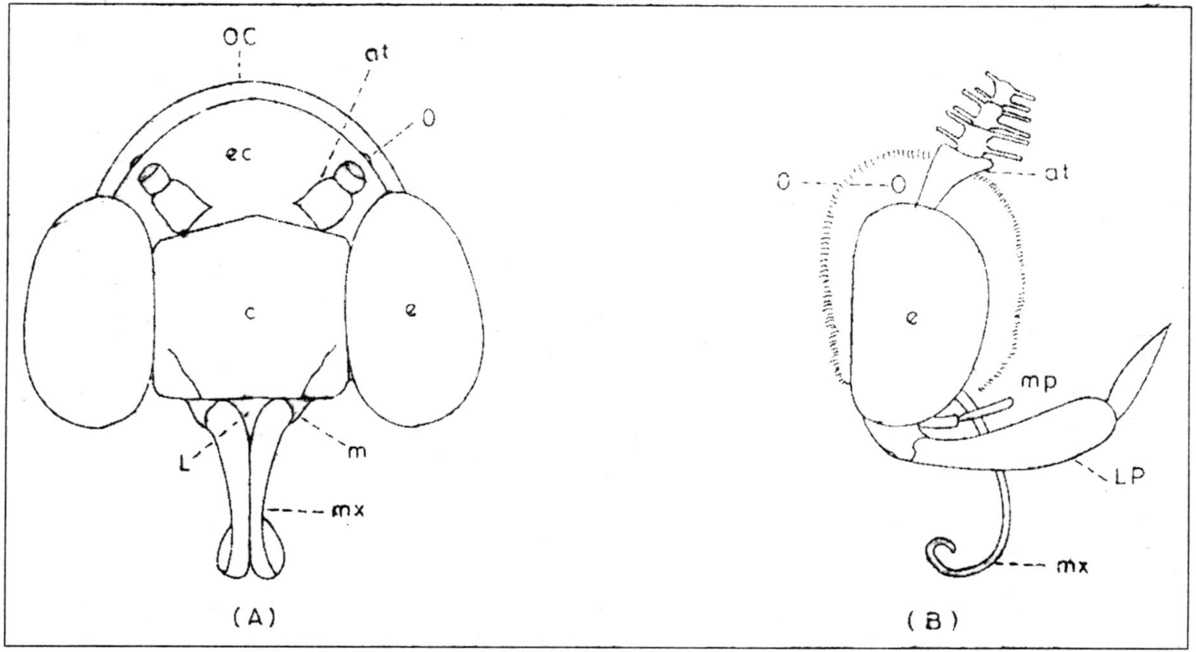

Figure 23: Morphology of Head; (A) Front view, (B): Side view
(A) OC: Occiput; ec: Epicranium; at: Antenna; c: Clypeus; e: Eye; m: Mandible; L: Labrum; O: Ocellus; mx: Proboscis
(B) LP: Labial palpi; mp: Maxillary palpi

with a great number of long stiff bristles which project over the base of the tongue. The genal process (2) is more or less triangular projection between the pilifer and tongue. Below the pilifer close to the tongue on each side is a short process which is the remnant of the maxillary palpus (3). It is densely clothed with long white scale. The transverse arched strip between the labial palpi is the mentun. The proboscis or tongue (4) is formed by the first pair of maxillae and consists of two halves closely applied to each other. The palpus is large broad in lateral aspect, closely contiguous to the head, and has a short mini segment.

Eyes are subglobular, never hairy, but often covered above by a kind of eye brow and below by a large tuft of hairs. Antennae is divisible into scape, pedicel and flagellum (Figure 24).

Thorax (Figure 25)

Mesonotum, composed of the prescutum, scutum and postscutum, is very large. The parascutum is distinctly triangular in dorsal view. The sentum (1) is widest behind and a little longer than broad. The post-scutum varies obviously in size and shape. Similar parts compose the metanotum. The scutum (3) is divided into two halves. The postscutum (3) is always narrow. The ventral pairs of the meso and metathorax do not differ much in size. The sternum and peristernum (12) are not completely separated from one other. The peristernum is large and remains broad at the obliquely truncate upper end where it leas against the parasternum (9). This is a large plate extending obliquely dorsad and mesiad from the meral suture (13) separating the meral and sternal parts of the sternite, to the membrane connecting meso and prothorax. Between this plate and the notum the mesothoracic tegular (7) is inserted. Below the parasternum there is episternum (11), with which are fused the hyposternum (15) and the marginal strips along the coxal cavity. The meral half of the sternite is made up of the paramerum (8) and the protomerum (13), two more or less strongly convex plates, together with the large epimerum (10). A marginal strip (14), situated along the meral cavity, is separated by a more or less distinct suture from, the epimerum. The metasternite is more simplified than the mesosternite.

Leg (Figure 27)

The leg comprises the coxa, trochanter, femur, tibia and tarsus.

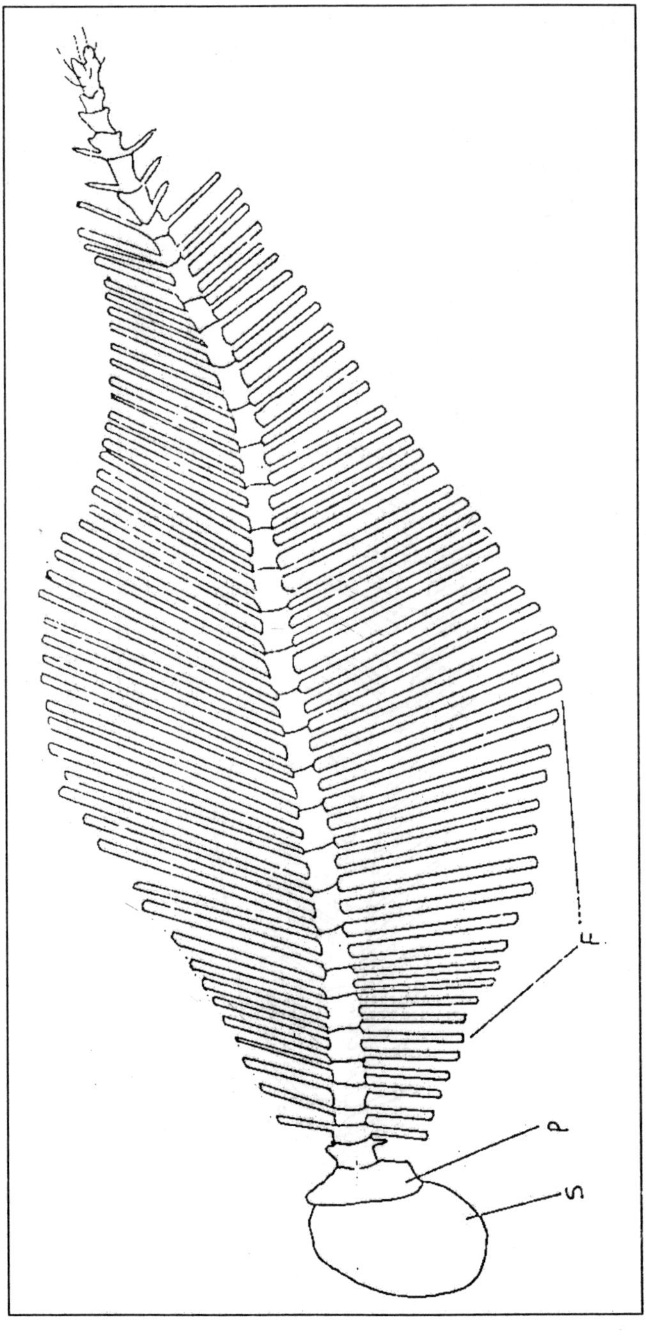

Figure 24A: Antennae of Male Silkmoth
P: Pedicel; S: Scape; F: Flagellum

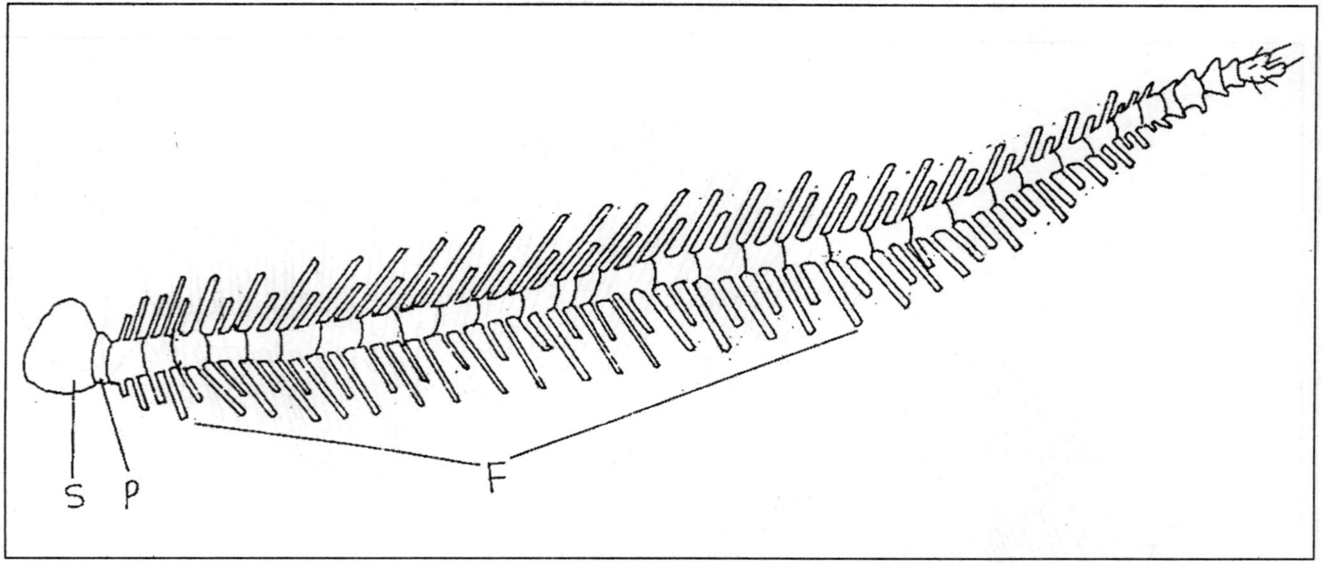

Figure 24B: Antennae of Female Silkmoth
P: Pedicel; S: Scape; F: Flagellum

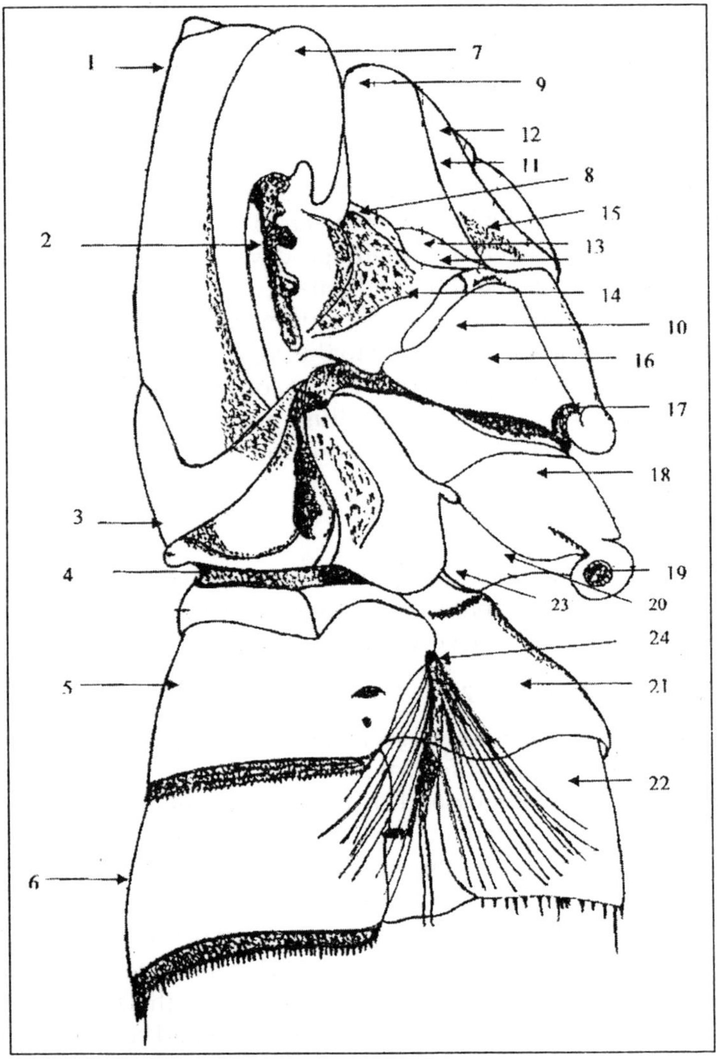

Figure 25: Morphology of Thorax with First Abdominal Segments

1: Scutum of mesothorax; 2: Inservation of wing; 3: Scutellum of mesothorax; 4, 5, 6: Tergites of first, second and third abdominal segments; 7: Mesothorax tegular; 8: Paramerum; 9: Parasternum; 10: Marginal strip; 11: Episternum; 12: Peristernum; 13: Paraplurum; 14: Marginal strip; 15: Hyposternum; 16, 20: Merum; 17: Trochanter; 18: Coxa; 19: Troch; 21, 22: Sternites of 2nd and 3rd abdominal segments; 23: Hypomerum

The coxa is inserted in a groove formed by sternal parts of the sternite. The trochanter (troch) is borned by the coxa and is supported behind by merum. The femora always remain simple. Tibia and tarsus may have several modifications. The apex of the fore tibia is often produced into a strong process (a thorn). Tibiae are more or less spinose. The mid tibia has one pair of slender spurs but, the proximal pair very often disappears. The hind tarsus is generally longer than the midtarsus. The comb is less strongly developed. The fifth segment of all tarsi bears some stout and pale sensory hairs at the end on each side close to the apical spine, forming often a brush. There are two long bristles dorsally close to the edge, curving ventrad. The claw segment is composed of the claw (onychium), the false claw (paranychium), the pad (pulvillus) and the empodium. The empodium is a small tubercle above the pad between the claws bearing one bristle.

Wings (Figure 26)

The veins of the wings are dealt with the systematic section. The frenulum and retinaculum are reduced/vestigial or absent. The fore and hind wings are very variable in shape.

Abdomen (Figure 22)

Moth has ten abdominal segments. The ninth and tenth of male and the eighth to tenth of female are modified. The spines stand at the edges of the segments and are found on segments 2 and 8 in the male and 2 and 7 in female. The first abdominal segments consists of a tergite (at) and a more or less triangular lateral plate, the parapleurum (pp), it bears no trace of real spines. The first abdominal stigma (sti) lies free in the membrane in front of the parapleurum. The second to sixth tergites are longer. The eighth tergite is small and partly (♂) or completely (♀) concealed by the seventh. The parapleura of segments 2 and 8 are membranous and bear the stigma. The sternites of the first and last segments may have modifications.

Genitalia (Figure 28)

The male genitalia consist of VII, VIII, IX and X abdominal tergites and sternites. Associated with these segments, there are periphallic organs, gnathos, saccus, uncus, harpes and accessory harpes including a phallus complex. The VII and VIII segments are

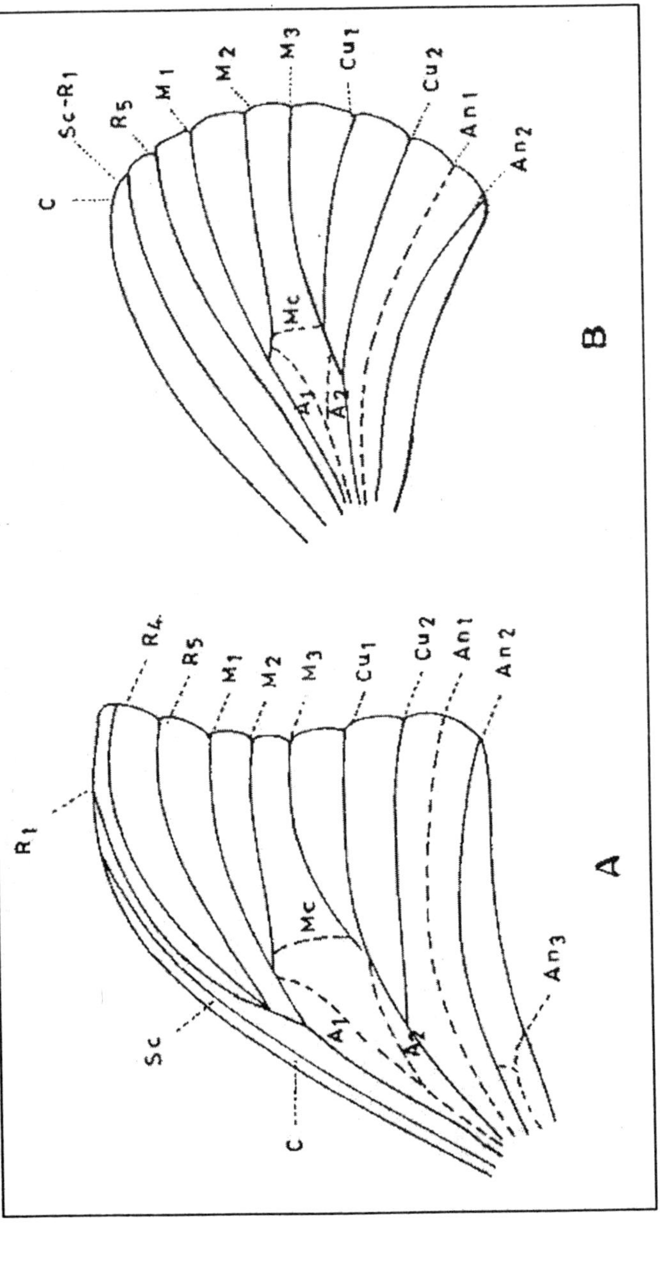

Figure 26: A: Fore wing; B: Hind wing (Diagrammatic)
C: Costa; Sc: Sub costa, Radius (R1, R2, R3); M: Media (m1-M2); Cu: Cubitus (Cu4, cu2);
An: Anal (An1, An2, An3); A: Weak vein (A1, A2), Mc: Median cross vein

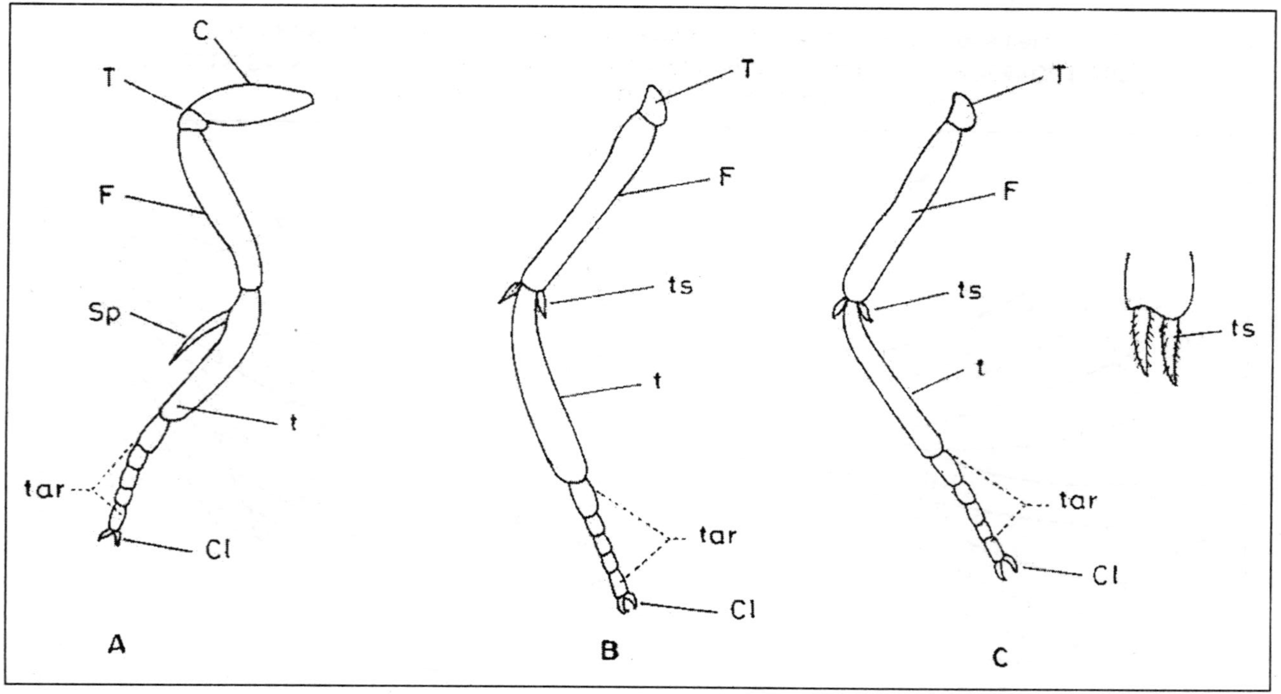

Figure 27: Leg of Moth
A: Fore wing–C: Coxa, T: Trochanter, F: Femur, T: Tibia, tar: Tarsus, S: Spur;
B: Middle leg, ts: Tibial spur, Cl: Claws; C: Hind leg, F: Femur

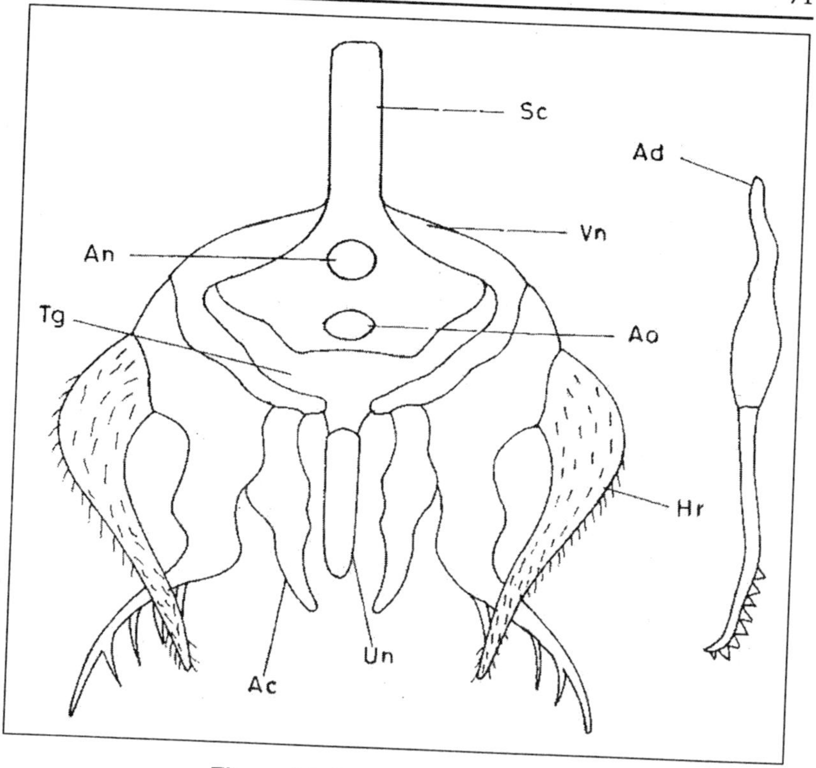

Figure 28: Structure of Genitalia
Sc: Saccus; Vn: Vinculum; An: Anellus; Ao: Anal opening; Hr: Harpes; Ac: Accessory harpes; Tg: Tegumen; Ad: Aedeagus

not modified in relation to the genitalia but form at least the protractile or retractile base for the copulatory apparatus and remain membranous. On the contrary, the IX and X segments are greatly modified and act as gonosomites. The IX gonosomite encircles the tip of lithe abdomen and forms the base of the peripheral genital processes. It exists in the form of a simple sclerite ring with distinct tergal and coxosternal areas of sclerotization. Mid-ventrally, its sternal portion (vinculum) is produced into a long median saccus which extends anteriorly upto the VII segment. The ventral wall of saccus is adnate with the VII and VIII sternites and strongly attached to them. The hind margin of the gonosomite (tegumen) there is a slender hook like median dorsal process called, uncus. It is projected posteriorly, the tip being attached to the hind margin of X tergum, which is membranous. A like ventral sclerite called gnathos (sub-anal structure) is hinged basally to uncus. It lies vertically, the serrated tip ending near the rudimentary anal opening of the male. A pair of movable genital claspers (harpes) hinged to the lateral sides of the vinculum forms the most prominent part of external genitalia. Each harpes is a tri lobed structure. The smallest lobe is situated in the ventro-lateral aspect of the vinculum. The middle lobe is the longest lies in its lateral aspect. The third lobe which is little smaller than the lateral lobe occupies a dorsolateral position. The first lobes are muscular and flexible bearing numerous hairs, whereas, the dorso-lateral lobe is chitinous and bears five spines at distal extremity. A chitinous accessory harpes, hinged to the vinculum on either side from below, the harpes occupies a position between the lateral and dorso-lateral lobes. It is broad at the base extending to about two third of its length, the rest being in the form of a slender curved needle.

The phallus complex of *A. mylitta* consists of an unpaired supporting structures (phallobase). The phallus includes adeagus, cornua, apodem and endophallus. The distal part of the phallus (adeagus} is a typical sclerotic-structure. It is slender with a curved and pointed tip. While its dorsal aspect is serrated, ventrally it is produced into a bow shaped plate. The middle portion of the phallus is swollen (cornua) having a small hole through which the ejaculatory duct passes into the endophallus.The apodem is a simple rod and forms the proximal part of the phallus. The phallobase is a mere inflection of the genital chamber wall forming a pocket (phallocrypt) containing the base of the adeagus and is produced as a tubular theca (phallotheca). The phallotheca is broader at the anterior side, slowly getting narrower posteriorly. Its posterior edge is attached to the proximal extremity of the adeagus, the anterior edge being fused with the anellus. The anellus is the sclerotic ring of the phallocrypt or phallotheca encircling the opening through which the adeagus passes. All these structures are associated with the mid-ventral portion of the vinculum. A pair of strong muscle bands arise each from cornua and apodem.

GENUS *ANTHERAEA* HUBNER

The genus *Antheraea* is erected by Hubner in 1818 on the basis of type, *A. paphia* Linn, Wood- Mason (1886) sub divided the genus *Antheraea* into two subgena *viz., Antheraea* and *Antheraeopsis.* The genus is scattered in Africa, Japan, China, Philippines, India, Sri Lanka, Myanmar, Andaman and Java. It is characterized by fore wing with Costa incurved near base, excurved towards apex, which is rounded; cell of both wings closed. From India Hampson (1892) reported seven species under this genus and recently Sathe (2007) confirmed seven species and he add one more new species under this genus from India. No subspecies have been described under the species *mylitta* of this genus upto date however, 44 ecoraces have been reported from India (Srivastava *et al.*, 2003).

KEY TO SUB SPECIES OF *ANTHERAEA MYLITTA*

1. Anterio lateral part of fore wing blunt
   Hyaline area more or less oval .... (3)
   Dorsolateral lobe with three spines
2. Anterio lateral part of fore wing acute or some what pointed
   Hyaline area not circular or oval .... (7)
   Dorsolateral lobe with two spines
3. Saccus anteriorly rounded and wider, slightly notched basally
   Adeagus knife shaped and black banded .... *A. mylitta indica*
   Dorsolateral lobe bears three long spines
4. Saccus anteriorly slightly tapering .... (5)
   Saccus not notched basally
   Dorsolateral lobe bears three long spines
   Adeagus long, knife shaped longer and broder than *indica* .... *A. mylitta jujubi*
5. Saccus very short, broder slightly tapering anteriorly
   Dorsolateral lobe bears three spines
   Adeagus long, straight, hockey stick shaped at the base and dumble shaped at anterior
   Genitalia more or less faint or light brown .... *A. mylitta arjuni*

6. Adeagus straight and pointed anterior .... (7)
   Saccus notched
   Moth appearance grayish ...... *A. mylitta grayei*
7. Inner spine blunt
   Adeagus longer, tapering anteriorly,
   Moth appearance not whitish ..... *A. mylitta badami*
8. Saccus longer, more or less rounded anteriorly
   Harpes broad, long, pointed
   Adeagus longer than genitalia
   excluding distal spines .. *A. mylitta kolhapuriensis*
9. Harpes very much broder, inner spine blunt
   Adeagus not straight .... (10)
   And equal to entire genitalia including
   distal spines .... *A. mylitta sathei*
10. Harpes longer and pointed
    Saccus notched
    Adeagus not banded and not straight ...... *A. mylitta koynei*
11. Saccus very narrow
    Dorsolateral spines very long
    Harpes broad, pointed and long
    Adeagus dark .... *A. mylitta sahydricus*

ANTHERAEA MYLITTA INDICA SUB SP. NOV

Male (Figure 29)

33 mm long, 8 mm broad, yellowish brown dorsally, brown ventrally; antenna light brown, bipectinate, 17.4 mm long, 7.1 mm broad, 37 segmented; fore leg 21.7 mm long, wing expanse 125mm.

Head

3mm long, 5 mm broad, brown dorsally, ventrally grey; eyes rounded, large, light brown in colour, ocular distance 15 mm; proboscis vestigial; antenna 17.4 mm long, 7.1 mm broad, 37 segmented, terminally bifurcated, yellowish brown, scape 1 mm long, pedicel 0.40 mm long, flagellum 16 mm long; labial palpi upturned, three segmented, brownish.

Flagellar Formula

1 L/W = 0.28, 15L/W= 1.75, T L/W= 2.4, A = 1.47

Thorax

10mm long, 8mm broad, dark brownish dorsally, greyish ventrally; pro, meso and metathorax brown with brown scales, scales 0.20 mm long.

Fore Wing (Figures 56a, 57a, 58a)

66 mm long, 33 mm broad, area of the fore wing 1251 mm^2, costal region of wing ash grey in colour; postmedian line pink with a white line on its border; antemedian line dark brown and bordered on the inside with a white line; oblique line brown with a indistinct faint white inner border; ventrally, pink, grey scales; antemedial line indistinct; medial line indistinct. Post medial line indistinct, margin in both wings brown. Ocellus (Figure 57a) 68 mm^2 with a hyaline area 6 mm^2, median cross vein straight and not touching with A1 nor A2, anterior outer line black, anterior inner line white, anterior half dove grey, posterior outer line black, posterior inner line light yellow, posterior half dove grey. Hyaline area oval; wing scales generally conical and bristle like with up to nine spines of different lengths. Anterio lateral edge blunt not pointed, fore wing outer curvature (Figure 58a) not 's' shaped, wing expanse – 125mm.

Hind Wing

43mm long, 35 mm broad, area of hind wing 1171mm^2; area of ocellus 62 mm^2, area of hyaline spot 4 mm^2; hyaline area oval shaped; antemedian line dark brown but narrow; oblique line absent, post median line prominent and broad.

Fore Leg

21.7 mm long, 1.6 mm broad, brown coloured; coxa 2.8 mm long; trochanter 1.0 mm long; femur 5.80 mm long; tibia 7.0 mm long, hind tibial spurs present; tarsus 4.5 mm long, five segmented; claw 0.60 mm long, curved, dark brown. scales 0.18mm long and 0.13mm wide.

Abdomen

20 mm long, 7 mm broad, dorsally brownish, densely covered with brownish scales on dorsal, ventral and lateral sides, scales 0.18mm long and 0.13mm wide.

Genitalia (Figure 30)

1. 10.10 mm long, 6.10 mm brcad, Uncus bifid, sparsely set with setae on the dorsal side, apex nothched, down curved, hook like median dorsal process, chitinised, ending into two pointed teeth on each side.
2. Vinculum very short, saccus short, slightly pointed, not rounded anteriorly and bulbus.
3. Harpes hinged to the lateral sides of vinculum, trilobed, first lobe muscular and flexible bearing numerous setae as well as strong bristles; dorsolateral lobe chitinous and bears three spines of distal extremity. Harpes spines comparatively larger.
4. Tegumen broder in the middle, narrow at the both ends, apically the end produced into a flattened process, latter broadened at its end.
5. Anellus very strong, chitinised in to a quadrate plate above which extends down words and inwards as conical prolongation.
6. Adeagus (Figure 31) 9.60 long, 0.85 mm broad, narrow, basal part shorter than apical part, latter denticulate in its distal on third part, ventrally produced into a bow shaped plate.

Cocoon

White grey in colour at surface, oval shaped, filament texture golden yellow.

1. Cocoon weight (g) : 8.42
2. Shell weight (g) : 1.12
3. Shell ratio (per cent) : 13.30
4. Cocoon length (cm) : 4.4
5. Cocoon width (cm) : 2.4
6. Peduncle length (cm) : 2.3
7. Peduncle width (mm) : 0.08
8. Peduncle weight (mg) : 0.10
9. Ring diameter (mm) : 7

10. Filament length (mt) : 292.20
11. Reeled weight (g) : 0.27
12. Denier : 8.31

Host Plants

Ber (*Z. jujuba*), Arjun (*T. arjuna*),

Ain (*T. tomentosa*), Badam (*T. catappa*).

Holotype

Male, India, Maharashtra, Coll. 18-VI-2008, Ajara (Ramthirth)

Kavane. R.P., leg, antenna, wing on card sheet, body pinned in insect box, labeled as above.

Paratype

12males: 3females, sex ratio (m: f) 4:1, coll, from May to December, same data as above.

Etymology

The sub sp. name *indica* refers to the found in India.

Distributional Record

Western Maharashtra, Ajara, Ramthirth 08-VII-2008; 1♂, Panhala 12-IV-2007; 1♂, Radhangari 23-VII-2007; 1♀,1♂, Patan 12-V-2007; 2♂, Anuskara 12-VI-2007; 1♂, Atigre 2-VIII-2007; 1♀,1♂, Dehu- alandi 28-V-2008; 1♂, Kokroud 12-VII-2007;2♂, Malakapur 12-V-2007; 1♂, Amba 12-V-2007; 1♀,1♂.

Remarks

Review of literature indicates that the present form runs close to *Antheraea mylitta* ecorace Sukinda by having following characters:

1.Cocoon weight 2) Shell weight 3) Silk ratio 4) Denier

However, it differs from above ecorace by having following characters.

1. Cocoon weight – 8.42 g
2. Shell weight – 1.12 g

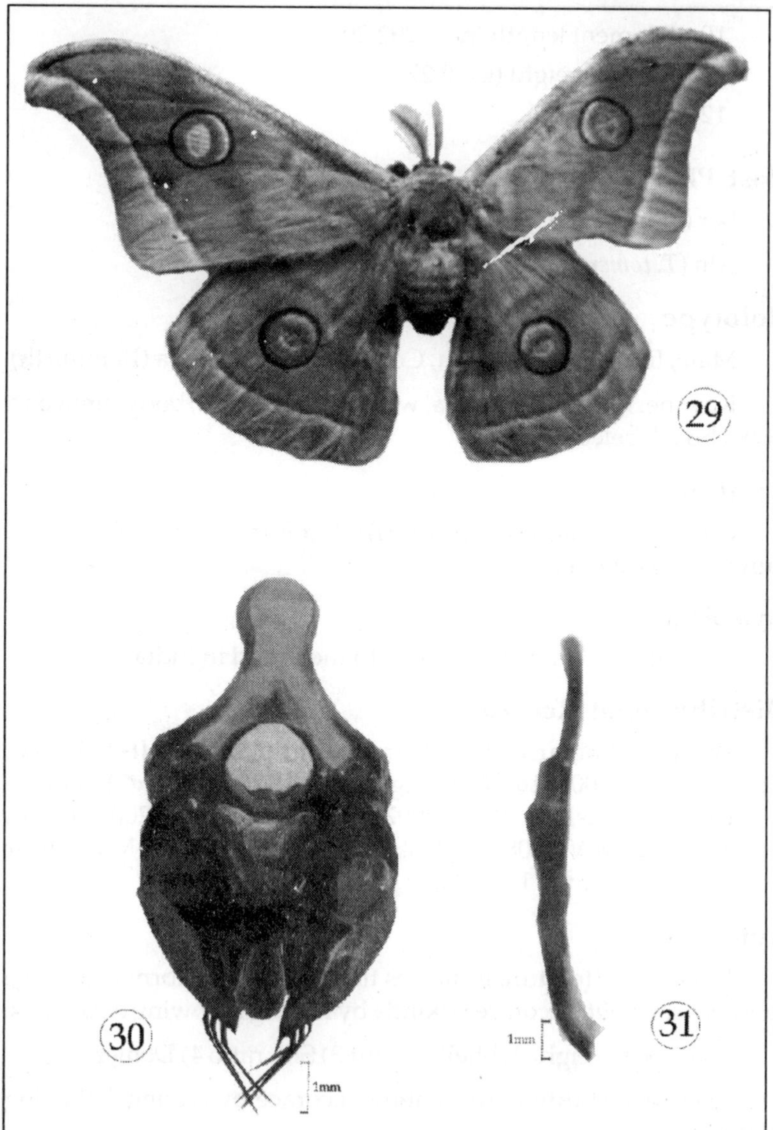

Plate 6

Figure 29: ***A. mylitta indica*–Adult male;**
Figure 30: ***A. mylitta indica*–Genitalia;**
Figure 31: ***A. mylitta indica–Adeagus***

3. Silk ratio – 13.30
4. Denier- 8.31
5. Morphological characters:
   (*a*) Hook like fore wing curvature
   (*b*) Medial line straight and strong
   (*c*) Hyaline area oval
   (*d*) Ocellus colour and radius
   (*e*) Genitalia size and shape

This sub species runs close to *jujubi* by having 3 spines on dorsolateral lobe. However, it differs by having following characters

1. Adeagus long, knife like, longer and broder than *indica*
2 Saccus anteriorly rounded and wider and slightly notched to base
3. Fore wing curvature
4. Hyaline area
5. Flagellar formula–1 L/W = 0.28, 15L/W= 1.75, T L/W= 2.4, A = 1.47

ANTHERAEA MYLITTA JUJUBI SUB SP.NOV

Male (Figure 32)

38 mm long, 12 mm broad, yellowish brown dorsally, brown ventrally; antenna light brown, except five apical segments, bipectinate, 17.4 mm long, 7.1 mm broad, 37 segmented; fore leg 21.7 mm long, wing expanse 138 mm.

Head

3mm long, 5 mm broad, brown dorsally, ventrally grey; eyes rounded, large, light brown in colour, ocular distance 15 mm; proboscis vestigial; antenna 17.4 mm long, 7.1 mm broad, 37 segmented, terminally bifurcated, yellowish brown, scape 1 mm long, pedicel 0.40 mm long, flagellum 16 mm long; labial palpi upturned, three segmented, brownish.

Flagellar Formula

1 L/W = 0.43, 15L/W= 1.39, T L/W= 1.76, A = 1.19

Thorax

12mm long, 10 mm broad, dark brownish dorsally, greyish ventrally; pro, meso and metathorax brown with brown scales, scales 0.20 mm long.

Fore Wing (Figures 56b, 57b, 58b)

73mm long, 49 mm broad, area of the fore wing 1295 mm^2, costal region of wing ash grey in colour; postmedian line pink with a white line on its border; antemedian line dark brown and bordered on the inside with a white line; oblique line brown with a indistinct faint white inner border; ventrally, pink, grey scales; antemedial line indistinct; medial line indistinct. Post medial line indistinct, margin in both wings brown. Ocellus (Figure 58b) 62mm^2 with a hyaline area 16 mm^2, median cross vein straight, anterior outer line reddish pink, anterior inner line white, anterior half dove grey, posterior outer line black, posterior inner line light yellow, posterior half dove grey. Hyaline area circular; wing scales generally conical and bristle like with up to nine spines of different lengths. Anterio lateral edge blunt not pointed, fore wing outer curvature(Figure 57b) not 'S' shaped, wing expanse – 138mm.

Hind Wing

49mm long, 45 mm broad, area of hind wing 922 mm^2; area of ocellus 52 mm^2, area of hyaline spot 6 mm^2; hyaline area oval shaped; antemedian line dark brown but narrow; oblique line absent, Post median line prominent and broad.

Fore Leg

21.7 mm long, 1.6 mm broad, brown coloured; coxa 2.8 mm long; trochanter 1.0 mm long; femur 5.80 mm long; tibia 7.0 mm long, hind tibial spurs present; tarsus 4.5 mm long, five segmented; claw 0.60 mm long, curved, dark brown.

Abdomen

23 mm long, 12 mm broad, dorsally brownish, densely covered with brownish scales on dorsal, ventral and lateral sides, scales 0.20 mm long,0.10 mm width.

Genitalia (Figure 33)

1. 11.20mm long, 6.10 mm broad, Uncus bifid, sparsely set with setae on the dorsal side, apex nothched, down curved,

hook like median dorsal process, chitinised, ending into two pointed teeth on each side.

2. Vinculum very short, saccus long slightly pointed, not rounded anteriorly and bulbus.
3. Harpes hinged to the lateral sides of vinculum, trilobed, first two lobes muscular and flexible bearing numerous hairs; median lobe more membranous, broad and densely set with setae; dorsolateral lobe chitinous and bears one short and two long spines at distal extremity, harpes spines comparatively larger and shorter.
4. Tegumen broder in the middle, narrow at the both ends, apically the end produced into a flattened process, latter broadened at its end.
5. Anellus very strong, chitinised, circular; anal opening chitinised into an almost quadrate plate.
6. Adeagus (Figure 34) 9.40 mm long, 0.85 mm broad, narrow, basal part broader than apical part, denticulate in its distal on third part, ventrally produced into a bow shaped plate.

Cocoon

Yellow in colour at surface, oval shaped, filament texture golden yellow.

1. Cocoon weight (g) : 9.42
2. Shell weight (g) : 0.95
3. Shell ratio (per cent) : 10.08
4. Cocoon length (cm) : 4.7
5. Cocoon width (cm) : 2.8
6. Peduncle length (cm) : 3.4
7. Peduncle width (mm) : 0.12
8. Peduncle weight (mg) : 0.12
9. Ring diameter (mm) : 9
10. Filament length (mt) : 282.30
11. Reeled weight (g) : 0.26
12. Denier : 8.28

Plate 7

Figure 32: ***A. mylitta jujubi*–Adult male;**
Figure 33: ***A. mylitta jujubi*–Genitalia;**
Figure 34: ***A. mylitta jujubi*–Adeagus**

Host Plants

Ber (Z. *jujuba*), Arjun (*T. arjuna*),

Asan (*T. tomentosa*), Badam (*T. catappa*)

Holotype

Male, India, Maharashtra, Coll. 8-VI-2008, Panhala, M.S,

Kavane R. P., antenna, leg, wing on card sheet, body pinned in insect box, labeled as above.

Paratype

12 males : 4 females, sex ratio (m:f) 3:1, coll, from May to December, same data as above.

Etymology

The sub sp. name *jujupi* refers to the collection on food plant ber.

Distributional Record

Western Maharashtra, Ramling 03-VII-2008; 1♀,1♂, Panhala 08-VI-2008; 1♂, Radhangari 23-VII-2008; 1♀,1♂, Patan 12-IV-2007; 2♂, Anuskara 12-IV-2008; 1♂, Atigre 2-IV-2009; 1♀,1♂, Phaltan 28-IV-2008; 1♂, Kokroud 12-IV-2006; 2♂, Shindewadi 12-IV-2007; 1♂, Shirala 12-IV-2007; 1♀,1♂.

Remarks

Review of literature indicates that the present form runs close to *Antheraea mylitta* ecorace Giribum by having following characters,

1) Cocoon weight 2) Shell weight 3) Silk ratio 4) Denier

However, it differs from above ecorace by having following characters.

1. Cocoon weight – 9.42 gm.
2. Shell weight – 0.95 gm.
3. Silk ratio – 10.08
4. Denier- 8.28
5. Morphological characters –
   (*a*) Hook like fore wing curvature
   (*b*) Medial line straight and strong

(*c*) Hyaline area circular

(*d*) Ocellus colour and radius

(*e*) Genitalia shape and size

- (*i*) Saccus on the top slightly pointed and not rounded anteriorly as in sub sp. *kolhapurensis.*
- (*ii*) Harpes spines comparatively larger than the sub sp. *kolhapurensis.*
- (*iii*) Distal spines relatively shorter than the sub sp. *kolhapurensis.*
- (*iv*) Adeagus shape and size.

(*f*) Anteriolateral edge blunt not pointed, fore wing curvature not 'S' shaped.

(*g*) Abdominal length, width proportion with sub sp. *kolhapurensis.*

This species runs close to *A.m. indica* by having three spines on dorsolateral lobe. However it differs from *indica* by having following characters

1. Genitalia more or less faint or brown
2. Adeagus long knife shaped longer and broder than *indica*
3. Saccus not notched basally
4. Harpes considerably longer and pointed
5. Fore wing curvature
6. Hyaline area
7. Flagellar formula – 1 L/W = 0.43, 15L/W= 1.39, T L/W= 1.76, A = 1.19

ANTHERAEA MYLITTA ARJUNI SUB SP.NOV

Male (Figure 35)

35mm long, 11 mm broad, yellowish brown dorsally, brown ventrally; antenna light brown,bipectinate, 17.45 mm long, 7.1 mm broad, 37 segmented; fore leg 21.7mm long, wing expanse 132mm.

Head

3mm long, 5 mm broad, brown dorsally, ventrally grey; eyes rounded, large, light brown in colour, ocular distance 15 mm;

proboscis vestigial; antenna 17.4 mm long, 7.1 mm broad, 37 segmented, terminally bifurcated, light brown, scape 1 mm long, pedicel 0.45 mm long, flagellum 16 mm long; labial palpi upturned, three segmented, brownish.

Flagellar Formula

1 L/W = 0.42, 15L/W= 1.32, T L/W= 2.87, A = 1.53

Thorax

11mm long, 8 mm broad, dark brownish dorsally, greyish ventrally; pro, meso and metathorax brown with brown scales, scales 0.15 mm long.

Fore Wing (Figures 56c, 57c, 58c)

73 mm long, 47 mm broad, area of the fore wing 1465 mm^2, costal region of wing ash grey in colour; postmedian line pink with a white line on its border; antemedian line dark brown and bordered on the inside with a white line; oblique line brown with a indistinct faint white inner border; ventrally, pink, grey scales; antemedial line indistinct; medial line indistinct. Post medial line indistinct, margin in both wings brown. Ocellus (Figure 58c) 66mm^2 with a hyaline area 10mm^2, median cross vein wavy and not touching with A1 norA2, anterior outer line reddish pink, anterior inner line white, anterior half pink, posterior outer line black, posterior inner line light yellow, posterior half dove grey. Hyaline area circular; wing scales generally conical and bristle like with up to nine spines of different length. Anterio lateral edge blunt not pointed, fore wing outer curvature(Figure 57c) not 'S' shaped, wing expanse – 132 mm.

Hind Wing

46 mm long, 40 mm broad, area of hind wing 932 mm^2; area of ocellus 52 mm^2, area of hyaline spot 6 mm^2; hyaline area oval shaped; antemedian line is dark brown; oblique line absent.

Fore Leg

21.7 mm long, 1.6 mm broad, brown coloured; coxa 2.8 mm long; trochanter 1.0 mm long; femur 5.80 mm long; tibia 7.0 mm long, tibial spurs present;tarsus 4.5 mm long, five segmented; claw 0.60 mm long, curved, dark brown.

Plate 8

Figure 35: ***A. mylitta arjuni*****–Adult male;**
Figure 36: ***A. mylitta arjuni*****–Genitalia;**
Figure 37: ***A.mylitta arjuni*****–Adeagus**

Abdomen

27 mm long, 11 mm broad, dorsally brownish, densely covered with brownish scales on dorsal, ventral and lateral sides, scales 0.15 mm long, 0.12 mm broad.

Genitalia (Figure 36)

1. 11.15 mm long, 6.10mm broad, Uncus bifid, sparsely set with setae on the dorsal side, apex nothched, down curved, hook like median dorsal process, chitinised, ending into two teeth on each side.
2. Vinculum very short, entering into a short and bulbus saccus.
3. Harpes hinged to the lateral sides of vinculum, trilobed, first lobes are muscular and flexible bearing numerous hairs; dorsolateral lobe chitinous and bears three spines of distal extremity.
4. Tegumen broder in the middle, narrow at the both ends, apically the end produced into a flattened process, latter broadened at its end.
5. Anellus very strong, chitinised, circular, anal opening chitinised into an almost quadrate plate.
6. Adeagus (Figure 37) 10.15mm long, 0.85mm broad, narrow, basal part shorter than apical part, latter denticulate in its distal on third part, ventrally produced into a bow shaped plate.

Cocoon

Bright yellow in colour at surface, oval shaped, filament texture golden yellow.

1. Cocoon weight (g) : 10.42
2. Shell weight (g) : 1.20
3. Shell ratio (per cent) : 11.51
4. Cocoon length (cm) : 5.1
5. Cocoon width (cm) : 2.8
6. Peduncle length (cm) : 2.6
7. Peduncle width (mm) : 20

8. Peduncle weight (mg) : 0.12
9. Ring diameter (mm) : 6
10. Filament length (mt) : 342.20
11. Reeled weight (g) : 0.42
12. Denier : 11.04

Host Plant

Ber (*Z. jujuba*), Arjun (*T. arjuna*), Ain (*T. tomentosa*).

Holotype

Male, India, Maharashtra, Coll. 18-VI-2008, Hatkanangle,

Kavane. R.P., leg, antenna, wing on card sheet, body pinned in insect box, labeled as above.

Paratype

10 males: 4 females, sex ratio (m:f) 1:0.4, coll, from May to December, same data as above.

Etymology

The sub sp. name *arjuni* refers to the collection on food plant arjuna.

Distributional Record

Western Maharashtra, Hatkanangle ramling 08-VII-2008; 1♀,1♂, Panhala 12-IV-2008; 1♂, Radhangari 23-VII-2008; 1♂, Patan 12-IV-2007; 1♀,1♂, Anuskara 18-IV-2008; 1♀,2♂, Atigre 2-IV-2008; 2♂, Dehu- Alandi˙28-IV-2007;1♂, Kokroud 12-IV-2008; 1♂, Malakapur 12-IV-2007; 1♀ .

Remarks

Review of literature indicates that the present form runs close to *Antheraea mylitta* ecorace Barharwa by having following characters,

1) Cocoon weight 2) Shell weight 3) Silk ratio 4) Denier

However, it differs from above ecorace by having following characters.

1. Cocoon weight – 10.42 gm.
2. Shell weight – 1.20 gm.

3. Silk ratio – 11.51
4. Denier- 11.04
5. Morphological characters –
   (*a*) Hook like fore wing curvature
   (*b*) Medial cross vein wavy and strong.
   (*c*) Hyaline area circular
   (*d*) Ocellus colour and radius
   (*e*) Genitalia shape and size

This species runs close to *A.mylitta* sub sp. *jujubi* by having three spines on dorsolateral lobe. However, it differs from above species by having following characters.

1. Adeagus long, straight, hockey stick shaped at base and dumble shaped anteriorly
2. Genitalia more or less faint or brown
3. Harpes very thick and long
4. Fore wing curvature
5. Hyaline area
6. Flagellar formula – 1 L/W = 0.42, 15L/W= 1.32, T L/W= 2.87, A = 1.53

ANTHERAEA MYLITTA GRAYI SUB SP.NOV

Male (Figure 38)

34mm long, 9 mm broad, yellowish brown dorsally, brown ventrally; antenna light brown,bipectinate, 18.7 mm long, 7 mm broad, 37 segmented; fore leg 22.00 mm long, wing expanse 126 mm.

Head

3mm long, 5 mm broad, brown dorsally, ventrally grey; eyes rounded, large, light brown in colour, ocular distance 15 mm; proboscis vestigial; antenna 18 mm long, 7 mm broad, 37 segmented, terminally bifurcated, light brown, scape 1 mm long, pedicel 0.40 mm long, flagellum 16 mm long; labial palpi upturned, three segmented, brownish.

Flagellar Formula

1 L/W = 0.33, 15L/W= 1. 5, T L/W= 1.66, A = 1.16

Thorax

11mm long, 8 mm broad, dark brownish dorsally, greyish ventrally; pro, meso and metathorax brown with brown scales, scales 0.18 mm long.

Fore Wing (Figures 56d, 57d, 58d)

62mm long, 39 mm broad, area of the fore wing 1265 mm^2, costal region of wing ash grey in colour; postmedian line pink with a white line on its border; antemedian line dark brown and bordered on the inside with a white line; oblique line brown with a indistinct faint white inner border; ventrally pink, grey scales; antemedial line indistinct; medial line indistinct. Post medial line indistinct, margin in both wings brown. Ocellus (Figure 58d) 65mm^2 with a hyaline area 13mm^2, median cross vein wavy and not touching with A1 norA2, anterior outer line reddish pink, anterior inner line white, anterior half dove grey, posterior outer line black, posterior inner line light yellow, posterior half dove grey. Hyaline area circular; wing scales generally conical and bristle like with up to nine spines of different lengths. Anterio lateral edge blunt not pointed, fore wing outer curvature (Figure 57d) not 'S' shaped, wing expanse – 126 mm.

Hind Wing

37mm long, 31 mm broad, area of hind wing 815 mm^2; area of ocellus 62 mm^2, area of hyaline spot 6 mm^2; hyaline area oval shaped; antemedian line dark brown but narrow; oblique line absent, post median line prominent and broad.

Fore Leg

22.00 mm long, 1.6 mm broad, brown coloured; coxa 2.8 mm long; trochanter 1.0 mm long; femur 5.80 mm long; tibia 7.0 mm long, tibial spurs present;tarsus 4.5 mm long, five segmented; claw 0.60 mm long, curved, dark brown.

Abdomen

20mm long, 9 mm broad, dorsally brownish, densely covered with brownish scales on dorsal, ventral and lateral sides, scales 0.18mm long, 0.15mm broad.

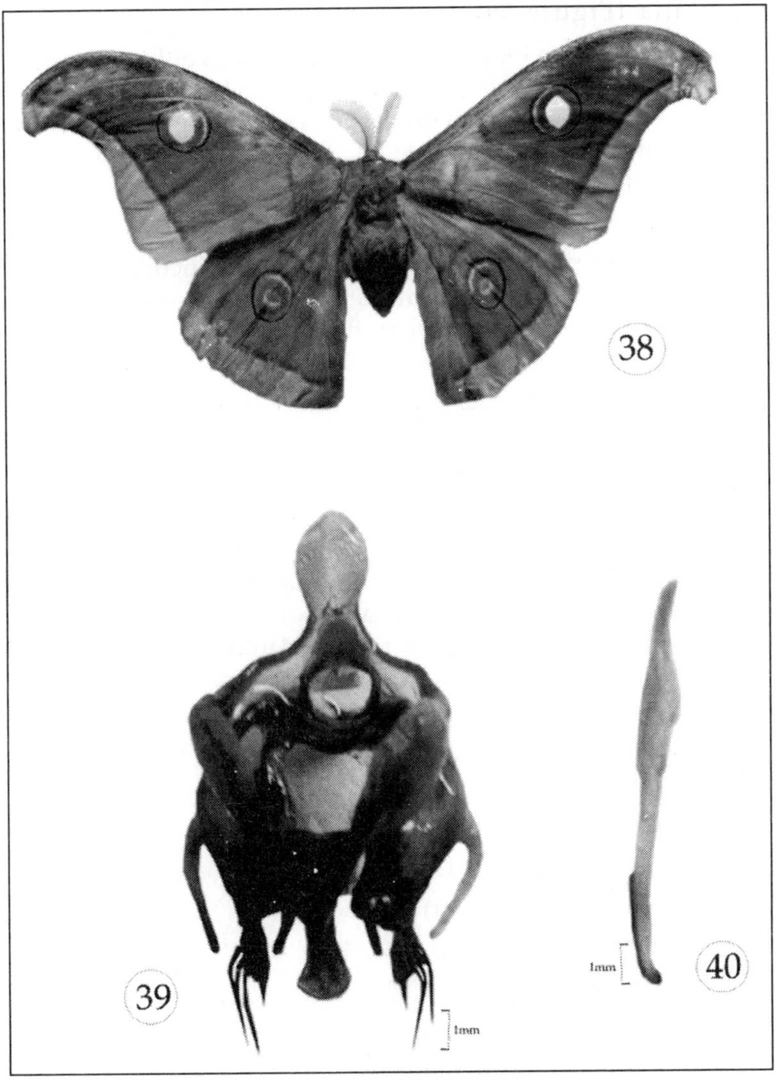

Plate 9

Figure 38: *A. mylitta greyi*–Adult male; Figure 39: *A. mylitta greyi*–Genitalia; Figure 40: *A. mylitta greyi*–Adeagus

Genitalia (Figure 39)

1. 11.80 mm long, 6.10 mm broad, Uncus bifid, sparsely set with setae on the dorsal side, apex nothched, down curved, hook like median dorsal process, chitinised, ending into two teeth on each side.
2. Vinculum very short, entering into a short, slightly pointed, not rounded anteriorly and bulbus saccus.
3. Harpes hinged to the lateral sides of vinculum, trilobed, first lobes muscular and flexible bearing hairs; dorsolateral lobe chitinous and bears three spines of distal extremity, harpes spine comparatively larger.
4. Tegumen broder in the middle, narrow at the both ends, apically the end produced into a flattened process, latter broadened at its end.
5. Anellus very strong, chitinised, circular; anal opening chitinised into an almost quadrate plate.
6. Adeagus (Figure 40) 9.10mm long, 0.85mm broad, narrow, basal part shorter than apical part, denticulate in its distal on third part, ventrally produced into a bow shaped plate.

Cocoon

White grey in colour at surface, oval shaped, filament texture golden yellow.

1. Cocoon weight (g) : 12.04
2. Shell weight (g) : 1.65
3. Shell ratio (per cent) : 13.70
4. Cocoon length (cm) : 4.6
5. Cocoon width (cm) : 2.8
6. Peduncle length (cm) : 2.4
7. Peduncle width (mm) : 14
8. Peduncle weight (mg) : 0.12
9. Ring diameter (mm) : 6
10. Filament length (mt) : 253.20
11. Reeled weight (g) : 0.45
12. Denier : 12.44

Host Plant

Ber (*Z. jujuba*), Arjun (*T. arjuna*), Ain (*T. tomentosa*).

Holotype

Male, India, Maharashtra, Coll. 8-VI-2008, Shindewadi,

Kavane. R.P., leg, antenna, wing on card sheet, body pinned in insect box, labeled as above.

Paratype

8males: 6females, sex ratio (m: f) 1:1.33, coll, from May to December, same data as above.

Etymology

The sub sp. name *grayi* refers to the colour of moth.

Distributional Record

Western Maharashtra, Hatkanangle ramling 03-VII-2008; 1♂, Panhala 12-IV-2007; 1♀,1♂, Radhangari 23-VII-2007; 1♀,1♂, Anuskara 12-IV-2008; 1♀,1♂, Atigre 2-IV-2008; 1♀,1♂, Shindewadi 8-IV-2008; 1♀,1♂, Kokrood 12-IV-2008; 1♂, Malakapur 12-IV-2007; 1♀, Amba 12-V-2007; 1♂.

Remarks

Review of literature indicates that the present form runs close to *Antheraea mylitta* ecorace Munga by having following characters,

1) Cocoon weight 2) Shell weight 3) Silk ratio 4) Denier

However, it differs from above ecorace by having following characters.

1. Cocoon weight – 12.04 gm.
2. Shell weight – 1.65 gm.
3. Silk ratio – 13.70
4. Denier- 12.44
5. Morphological characters –
   (*a*) Hook like fore wing curvature
   (*b*) Medial cross vein wavy and strong
   (*c*) Hyaline area circular

(*d*) Ocellus colour and radius

(*e*) Genitalia shape and size

This species runs close to *A.mylitta indica* by having three spines of dorsolateral lobe and ocellus. However, it differ from *indica* by having following characters.

1. Saccus notched
2. Moth appearance grayish
3. Adeagus straight
4. Entire genitalia dark except saccus
5. Fore wing curvature
6. Hyaline area
7. Antemedian line somewhat zigzag in fashion
8. Flagellar formula – 1 L/W = 0.33, 15L/W= 1. 5, T L/W= 1.66, A = 1.16

ANTHERAEA MYLITTA KOLHAPURENSIS SUB.SP.NOV

Male (Figure 41)

33 mm long, 9 mm broad, yellowish brown dorsally, brown ventrally; antenna light brown, except a five apical segments, bipectinate, 17.8 mm long, 7.2 mm broad, 37 segmented; fore leg 21.5 mm long, wing expanse 144mm.

Head

3 mm long, 5 mm broad, brown dorsally, ventrally grey; eyes rounded, large, light brown in colour, ocular distance 15 mm; proboscis vestigial; antenna 17.8 mm long, 7.2 mm broad, 37 segmented, terminally bifurcated, yellowish brown, scape 1 mm long, pedicel 0.40 mm long, flagellum 16 mm long; labial palpi upturned, three segmented, brownish.

Flagellar Formula

1 L/W = 0.33, 15L/W= 1.52, T L/W= 1.75, A = 1.20

Thorax

11mm long, 8 mm broad, dark brownish dorsally, greyish

ventrally; pro, meso and metathorax brown with brown scales; scales 0.20 mm long.

Fore Wing (Figures 56e, 57e, 58e)

69 mm long, 50 mm broad, area of the fore wing 1380 mm^2, costal region of wing ash grey in colour; postmedian line pink with a white line on its border; antemedian line dark brown and bordered on the inside with a white line; oblique line brown with a indistinct faint white inner border; ventrally, pink, grey scales; antemedial line indistinct; medial line indistinct. Post medial line indistinct, margin in both wings brown. Ocellus (Figure 58e) 73mm^2 with a hyaline area 27 mm^2, median cross vein concave and not touching with A1 nor A2, anterior outer line reddish pink, anterior inner line white, anterior half dove grey, posterior outer line black. Posterior inner line light yellow, posterior half dove grey. Hyaline area oval; wing scales generally conical and bristle like with up to nine spines of different lengths. Anterio lateral edge blunt not pointed, fore wing outer curvature (Figure 57e) not 'S' shaped. Wing expanse – 144 mm

Hind Wing

40 mm long, 45 mm broad, area of hind wing 945 mm^2; area of ocellus 57 mm^2, area of hyaline spot 6 mm^2; hyaline area oval shaped; antemedian line is dark brown; oblique line absent.

Fore Leg

21.5 mm long, 1.6 mm broad, brown coloured; coxa 2.8 mm long; trochanter 1.0 mm long; femur 5.80 mm long; tibia 7.0 mm long, Hind tibial spurs present;tarsus 4.5 mm long, five segmented; claw 0.60 mm long, curved, dark brown.

Abdomen

19 mm long, 9 mm broad, dorsally brownish, densely covered with brownish scales on dorsal, ventral and lateral sides. Scales 0.20 mm long and 0.15 mm broad.

Genitalia (Figure 42)

1. 11.20 mm long, 5.10 mm broad, Uncus bifid, sparsely set with setae on the dorsal side, apex nothched, down curved, hook like median dorsal process, chitinised, ending into two pointed teeth on each side.

2. Vinculum very short, saccus long and rounded at its end.
3. Harpes hinged to the lateral sides of vinculum, trilobed, first lobes are muscular and flexible bearing numerous hairs; dorsolateral lobe chitinous and bears two spines of distal extremity.
4. Tegumen broder in the middle, narrow at the both ends, apically the end produced into a flattened process, latter broadened at its end.
5. Anellus strong, chitinised, circular, anal opening chitinised into an almost quadrate plate.
6. Adeagus (Figure 43) 9.80 mm long, 0.85mm broad, narrow, basal part shorter than apical part, latter denticulate in its distal on third part, ejaculatory duct enters at the side near the base, the distal end slightly bent and serrated.

Cocoon

Bright yellow in colour at surface, oval shaped, filament texture golden yellow.

1. Cocoon weight (g) : 9.35
2. Shell weight (g) : 1.40
3. Shell ratio (per cent) : 14.97
4. Cocoon length (cm) : 3.6
5. Cocoon width (cm) : 2.3
6. Peduncle length (cm) : 3.6
7. Peduncle width (mm) : 23
8. Peduncle weight (mg) : 0.10
9. Ring diameter (mm) : 4
10. Filament length (mt) : 265.40
11. Reeled weight (g) : 0.26
12. Denier : 8.81

Host Plants

Ber (*Z. jujuba*), Arjun (*T. arjuna*),

Desi badam (*T. catappa*), Asan (*T. tomentosa*).

Plate 10

Figure 41: *A. mylitta kolhapurensis*–Adult male; Figure 42: *A. mylitta kolhapurensis*–Genitalia; Figure 43: *A. mylitta kolhapurensis*–Adeagus

Holotype

Male, India, Maharashtra, Coll. 8-VI-2007, Hatkanangle (Ramling), Kavane. R.P., leg, antenna, wing on card sheet, body pinned in insect box, labeled as above.

Paratype

12 males: 24 females, sex ratio (m: f) 2:1, coll, from May to December, same data as above.

Etymology

The sub sp. name *kolhapurensis* refers to the found in Kolhapur.

Distributional Record

Western Maharashtra, Hatkanangle (Ramling) 03-VII-2008; 3♀,1♂, Panhala 12-IV-2007; 3♀,1♂, Radhangari 23-VII-2007; 3♀,1♂, Patan 12-IV-2007; 3♀,2♂, Anuskara 12-IV-2007; 1♀,1♂, Atigre 2-IV-2007; 3♀,1♂, Dehu- Alandi 28-IV-2008; 3♀,1♂, Kokrood 12-IV-2007; 2♀,2♂, Malakapur 12-IV-2007; 2♀,1♂, Amba 12-IV-2007; 2♀,1♂.

Remarks

Review of literature indicates that the present form runs close to *Antheraea mylitta* ecorace Sarihan by having following characters,

1) Cocoon weight 2) Shell weight 3) Silk ratio 4) Denier

However, it differs from above ecorace by having following characters.

1. Cocoon weight – 9.35 gm.
2. Shell weight – 1.40 gm.
3. Silk ratio – 14.97.
4. Denier- 8.81
5. Morphological characters –
   (*a*) Hook like fore wing curvature
   (*b*) Medial line convex and strong
   (*c*) Hyaline area–Oval shaped
   (*d*) Ocellus colour and radius
   (*e*) Genitalia shape and size

This sub species runs close to *A. mylitta koynei* by having two spines on dorsolateral lobe and having blackish brown colour to genitalia. However, it differs from above species by following characters.

1. Proportion of width of genitalia and length of harper, in *A. mylitta kolhapurensis* is breadth is more than the *koynei* and length of harper is smaller than koynayi. In *A. mylitta kolhapurensis* saccus is more or less parallel sided and in *A.mylitta koynei* it is completely notched.
2. Shape of adeagus and its colour
3. Fore wing curvature
4. Hyaline area
5. Antemedian line
6. Saccus longer more or less rounded anteriorly
7. Harpes broad long. Pointed
8. Adeagus longer than genitalia excluding distal spines
9. Flagellar formula – 1 L/W = 0.33, 15L/W= 1.52, T L/W= 1.75, A = 1.20

ANTHERAEA MYLITTA KOYNEI SUB SP.NOV

Male (Figure 44)

34 mm long, 11 mm broad, yellowish brown dorsally, brown ventrally; antenna light brown,bipectinate, 17.9 mm long, 7.3 mm broad, 37 segmented; fore leg 21.9 mm long, wing expanse 125 mm.

Head

3 mm long, 5 mm broad, brown dorsally, ventrally grey; eyes rounded, large, light brown in colour, ocular distance 15 mm; proboscis vestigial; antenna 17.9 mm long, 7.3 mm broad, 37 segmented, terminally bifurcated, yellowish brown, scape 1 mm long, pedicel 0.40 mm long, flagellum 16 mm long; labial palpi upturned, three segmented, brownish.

Flagellar Formula

1 L/W = 0.40, 15L/W= 1.33, T L/W= 1.81, A = 1.18

Thorax

11mm long, 11 mm broad, dark brownish dorsally, greyish ventrally, pro, meso and metathorax brown with brown scales, scales 20 mm long.

Fore Wing (Figures 56f, 57f, 58f)

60 mm long, 37 mm broad, area of the fore wing 1365 mm^2, costal region of wing ash grey in colour; postmedian line pink with a white line on its border; antemedian line dark brown and bordered on the inside with a white line; oblique line brown with a indistinct faint white inner border; ventrally, pink, grey scales; antemedial line indistinct; medial line indistinct. Post medial line indistinct, margin in both wings brown. Ocellus (Figure 58f) 82 mm^2 with a hyaline area 25 mm^2, median cross vein straight and not touching with A1 nor A2, anterior outer line reddish pink, anterior inner line white, anterior half dove grey, posterior outer line black, posterior inner line light yellow, posterior half dove grey. Hyaline area circular; wing scales generally conical and bristle like with up to nine spines of different lengths. Anterio lateral edge blunt not pointed, fore wing outer curvature (Figure 57f) not 'S' shaped, wing expanse – 125 mm.

Hind Wing

36 mm long, 25 mm broad, area of hind wing 922 mm^2; area of ocellus 52 mm^2, area of hyaline spot 4 mm^2; hyaline area oval shaped; antemedian line is dark brown; oblique line absent.

Fore Leg

21.7 mm long, 1.5 mm broad, brown coloured; coxa 2.8 mm long; trochanter 1.0 mm long; femur 5.80 mm long; tibia 7.0 mm long, tibial spurs present;tarsus 4.5 mm long, five segmented; claw 0.60 mm long, curved, dark brown.

Abdomen

20 mm long, 10 mm broad, dorsally brownish, densely covered with brownish scales on dorsal, ventral and lateral sides, Scales 0.20 mm long and 0.18 mm broad.

Genitalia (Figure 45)

1. 11.90 mm long, 5.40 mm broad, Uncus bifid, sparsely set with setae on the dorsal side, apex nothched, down curved,

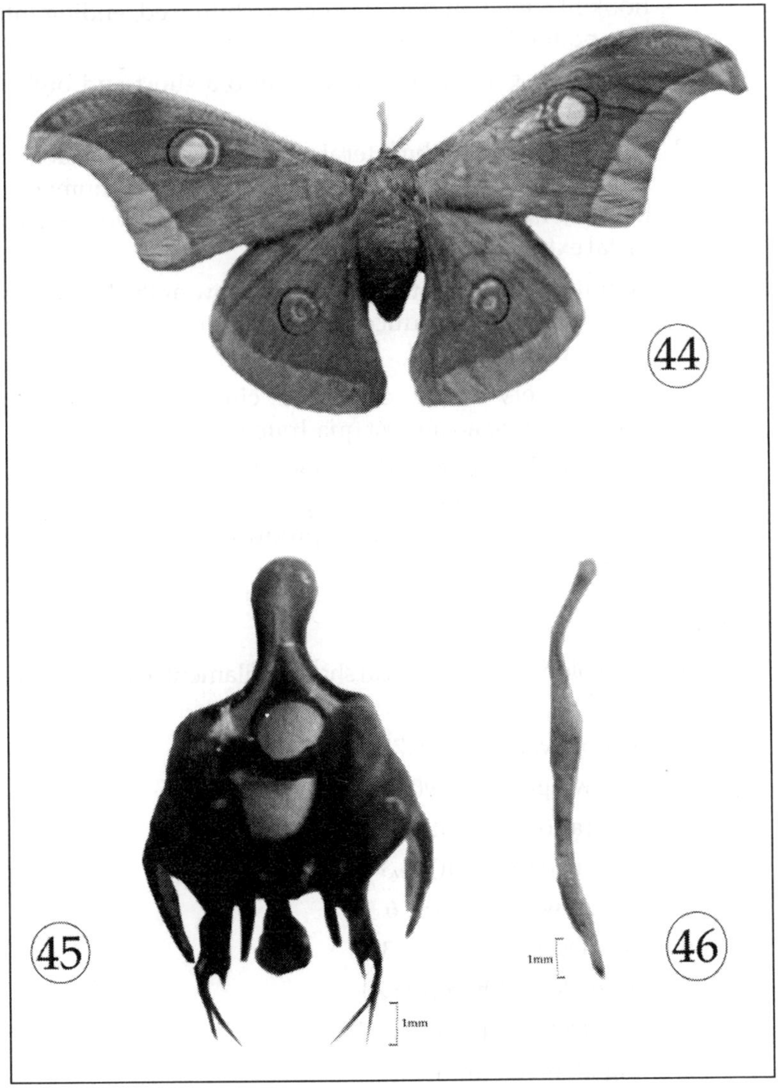

Plate 11

Figure 44: *A. mylitta koyanei*–Adult male;
Figure 45: *A. mylitta koyanei*–Genitalia;
Figure 46: *A. mylitta koyanei*–Adeagus

hook like median dorsal process, chitinised, ending into two pointed teeth on each side.

2. Vinculum very short, entering into a short and bulbus saccus.
3. Harpes hinged to the lateral sides of vinculum, trilobed, first two lobes are muscular and flexible bearing numerous hairs; dorsolateral lobe chitinous and bears two spines of distal extremity.
4. Tegumen broder in the middle, narrow at the both ends, apically the end produced into a flattened process, latter broadened at its end.
5. Anellus very strong, chitinised, circular, anal opening chitinised into an almost quadrate plate.
6. Adeagus (Figure 46) 9.85 mm long, 0.85mm broad, narrow, basal part shorter than apical part, latter denticulate in its distal on third part, ventrally produced into a bow shaped plate.

Cocoon

Yellow in colour at surface, oval shaped, filament texture golden yellow.

1. Cocoon weight (g) : 8.75
2. Shell weight (g) : 0.90
3. Shell ratio (per cent) : 10.28
4. Cocoon length (cm) : 4.3
5. Cocoon width (cm) : 2.6
6. Peduncle length (cm) : 3.2
7. Peduncle width (mm) : 20
8. Peduncle weight (mg) : 0.10
9. Ring diameter (mm) : 9
10. Filament length (mt) : 245.00
11. Reeled weight (g) : 0.20
12. Denier : 7.34

Host Plant

Ber (*Z. jujuba*), Arjun (*T. arjuna*).

Holotype

Male, India, Maharashtra, Coll. 18-VI-2008, Patan.

Kavane. R.P., leg, antenna, wing on card sheet, body pinned in insect box, labeled as above.

Paratype

08 males: 04 females, sex ratio (m:f) 2:1, coll, from May to December, same data as above.

Etymology

The sub sp. name *koynei* refers to the found in Koyna region Patan.

Distributional Record

Western Maharashtra, Hatkanangle 08-VII-2008; 1♂, Panhala 12-IV-2007; 1♀,1♂, Radhangari 23-VII-2007; 1♂, Patan 12-V-2007; 2♂, Anuskara 12-VI-2007; 1♀,Atigre 2-VI-2007; 1♂, Dehu-Alandi 28-VI-2008; 1♀, Kokroud 12-VII-2008; 1♂, Malakapur 12-VI-2007; 1♂, Amba 12-VIII-2007; 1♀.

Remarks

Review of literature indicates that the present form runs close to *Antheraea mylitta* ecorace Nowgaon by having following characters,

1) Cocoon weight 2) Shell weight 3) Silk ratio 4) Denier

However, it differs from above ecorace by having following characters.

1. Cocoon weight – 8.75 gm.
2. Shell weight – 0.90 gm.
3. Silk ratio – 10.28
4. Denier- 7.34
5. Morphological characters –
   (*a*) Hook like fore wing curvature
   (*b*) Medial cross vein straight and strong
   (*c*) Hyaline area circular
   (*d*) Ocellus colour and radius
   (*e*) Genetilia shape and size

This sub species runs close to *A. mylitta kolhapurensis* by having two spines on dorsolateral lobe and having blackish brown colouration of genitalia. However, it differs from *A. mylitta kolhapurensis* by following characters.

1. Harpes longer and pointed
2. Saccus notched
3. Adeagus not banded, not straight, colour is faint yellow and not dark brown
4. Fore wing curvature
5. Antemedial line
6. Proportion of genitalia width and harper length
7. Flagellar formula – 1 L/W = 0.40, 15L/W= 1.33, T L/W= 1.81, A = 1.18

ANTHERAEA MYLITTA SATHEI SUB SP.NOV

Male (Figure 47)

35 mm long, 11 mm broad, yellowish brown dorsally, brown ventrally; antenna yellowish brown,bipectinate, 17 mm long, 7 mm broad, 37 segmented; fore leg 21.4 mm long, wing expanse 125mm

Head

3 mm long, 5 mm broad, brown dorsally, ventrally grey; eyes rounded, large, light brown in colour, ocular distance 15 mm; proboscis vestigial; antenna 17 mm long, 7 mm broad, 37 segmented, terminally bifurcated, light brown, scape 1 mm long, pedicel 0.42 mm long, flagellum 18 mm long; labial palpi upturned, three segmented, brownish.

Flagellar Formula

1 L/W = 0.36, 15L/W= 1.32, T L/W= 2.87, A = 1.53

Thorax

11mm long, 8 mm broad, dark brownish dorsally, greyish ventrally; pro, meso and metathorax brown with brown scales, scales 20 mm long.

Fore Wing (Figures 56g, 57g, 58g)

62 mm long, 42 mm broad, area of the fore wing is about 1268

mm^2, costal region of wing ash grey in colour; postmedian line pink with a white line on its border; antemedian line dark brown and bordered on the inside with a white line; oblique line brown with a indistinct faint white inner border; ventrally, pink, grey scales; antemedial line indistinct; medial line indistinct. Post medial line indistinct, margin in both wings brown. Ocellus (Figure 58g) 72 mm^2 with a hyaline area 24 mm^2, median cross vein straight and not touching with A1 norA2, anterior outer line reddish pink, anterior inner line white, anterior half dove grey, posterior outer line black, posterior inner line light yellow, posterior half dove grey. Hyaline area circular; wing scales generally conical and bristle like with spines of different lengths. Anterio lateral edge blunt not pointed, fore wing outer curvature (Figure 57g) not 'S' shaped. Hook like curvature of fore wing, wing expanse –125 mm.

Hind Wing

38 mm long, 34 mm broad, area of hind wing 785 mm^2; area of ocellus 68 mm^2, area of hyaline spot 6 mm^2; hyaline area oval shaped; antemedian line is dark brown; oblique line absent.

Fore Leg

21.4 mm long, 1.5 mm broad, brown coloured; coxa 2.8 mm long; trochanter 1.0 mm long; femur 5.80 mm long; tibia 7.0 mm long, tibial spurs present;tarsus 4.5 mm long, five segmented; claw 0.60 mm long, curved, dark brown.

Abdomen

21mm long, 10 mm broad, dorsally brownish, densely covered with brownish scales on dorsal, ventral and lateral sides. Scales 0.20 mm long and 0.18 mm broad.

Genitalia (Figure 48)

1. 11.20 mm long, 6.10 mm broad, Uncus bifid, sparsely set with setae on the dorsal side, apex nothched, down curved, hook like median dorsal process, chitinised, ending into two teeth on each side.
2. Vinculum short, entering into a short and bulbus saccus.
3. Harpes hinged to the lateral sides of vinculum, first lobe muscular and flexible bearing numerous hairs; dorsolateral lobe chitinous and bears two spines of distal extremity.

Plate 12

Figure 47: *A. mylitta sathei*–Adult male;
Figure 48: *A. mylitta sathei*–Genitalia;
Figure 49: *A. mylitta sathei*–Adeagus

4. Tegumen broder in the middle, narrow at the both ends, apically the end produced into a flattened process, latter broadened at its end.
5. Anellus very strong, chitinised, circular, anal opening chitinised into an almost Quadrate plate.
6. Adeagus (Figure 49) 10.10 mm long, 0.85mm broad, narrow, basal part shorter than apical part, latter denticulate in its distal on third part, ventrally produced plate.

Cocoon

Bright yellow in colour at surface, oval shaped, filament texture golden yellow.

1. Cocoon weight (g) : 11.42
2. Shell weight (g) : 1.65
3. Shell ratio (per cent) : 14.44
4. Cocoon length (cm) : 4.2
5. Cocoon width (cm) : 2.7
6. Peduncle length (cm) : 2.4
7. Peduncle width (mm) : 20
8. Peduncle weight (mg) : 0.11
9. Ring diameter (mm) : 5
10. Filament length (mt) : 285.30
11. Reeled weight (g) : 0.44
12. Denier : 13.88

Host Plant

Ber (*Z. jujuba*), Arjun (*T. arjuna*), Ain (*T. tomentosa*).

Holotype

Male, India, Maharashtra, Coll. 8-VI-2008, Hatkanangle.

Kavane. R.P., leg, antenna, wing on card sheet, body pinned in insect box, labeled as above.

Paratype

10 males: 4 females, sex ratio (m:f) 1:0.4, coll, from May to December. Same data as above.

Etymology

The sub sp. name *sathei* refers to the honour of research guide.

Distributional Record

Western Maharashtra, Hatkanangle 03-VII-2008; 1♀,2♂, Panhala 12-IV-2008; 1♀,1♂, Radhangari 28-VII-2007; 1♀,1♂, Patan 12-IV-2007; 0♀,1♂, Anuskara 18-IV-2008; 1♀,1♂, Dehu- Alandi 28-IV-2007; 0♀,1♂, Kokrood 12-IV-2008; 0♀,1♂, Malakapur 12-IV-2007; 0♀,2♂.

Remarks

Review of literature indicates that the present form runs close to *Antheraea mylitta* ecorace Sukinda by having following characters,

1) Cocoon weight 2) Shell weight 3) Silk ratio 4) Denier

However, it differs from above ecorace by having following characters.

1. Cocoon weight – 11.42 gm.
2. Shell weight – 1.65 gm.
3. Silk ratio – 14.44
4. Denier – 13.89
5. Morphological characters –
   (*a*) Hook like fore wing curvature
   (*b*) Medial cross vein straight and strong
   (*c*) Hyaline area circular
   (*d*) Ocellus colour and radius
   (*e*) Genitalia shape and size

This sub species *A. mylitta sathei* runs close to *A. mylitta kolhapurensis* by having two spines on dorsolateral lobe and anterio lateral fore wing acute or some what pointed, hyaline area not circular. However, it differs from *kolhapurensis* by following characters.

1 Harpes very much broder
2. Inner spine blunt
3. Adeagus not straight
4. Entire genitalia light, yellow in colour except the inner third lobes.

5. Saccus narrowing anteriorly
6. Fore wing curvature
7. Antemedial line
8. Hyaline area
9. Flagellar formula– 1 L/W = 0.36, 15L/W= 1.32, T L/W= 2.87, A = 1.53

ANTHERAEA MYLITTA BADAMI SUB SP.NOV

Male (Figure 50)

31 mm long, 8mm broad, yellowish brown dorsally, brown ventrally; antenna yellowish brown,bipectinate, 17.6 mm long, 7.3 mm broad, 37 segmented; fore leg 21.5 mm long, wing expanse 126 mm.

Head

3 mm long, 5 mm broad, brown dorsally, ventrally grey; eyes rounded, large, light brown in colour, ocular distance 15 mm; proboscis vestigial; antenna 17.6 mm long, 7.3 mm broad, 37 segmented, terminally bifurcated, yellowish brown, scape 1 mm long, pedicel 0.40 mm long, flagellum 14 mm long; labial palpi upturned, three segmented, brownish.

Flagellar Formula

1 L/W = 0.33, 15L/W= 1.50, T L/W= 1.25, A = 1.02

Thorax

10 mm long, 8 mm broad, dark brownish dorsally, greyish ventrally, pro, meso and metathorax brown with brown scales, scales 20 mm long.

Fore Wing (Figures 56h, 57h, 58h)

62 mm long, 37 mm broad, area of the fore wing 1223 mm^2, costal region of wing ash grey in colour; postmedian line pink with a white line on its border; antemedian line dark brown and bordered on the inside with a white line; oblique line brown with a indistinct faint white inner border; ventrally, pink, grey scales; antemedial line indistinct; medial line indistinct. Post medial line indistinct, margin in both wings brown. Ocellus (Figure 58h) 67mm^2 with a hyaline area 7 mm^2, median cross vein concave and not touching

with A1 nor A2, anterior outer line reddish pink, anterior inner line white, anterior half dove grey, posterior outer line black. Posterior inner line light yellow, posterior half dove grey, hyaline area oval; wing scales generally conical and bristle like with up to nine spines of different lengths. Anterio lateral edge blunt not pointed, fore wing outer curvature (Figure 57h) not 'S' shaped. Wing expanse – 126mm.

Hind Wing

39 mm long, 32 mm broad, area of hind wing 782 mm^2; area of ocellus 64 mm^2, area of hyaline spot 4 mm^2; hyaline area oval shaped; antemedian line is dark brown; oblique line absent.

Fore Leg

21.5 mm long, 1.5 mm broad, brown coloured; coxa 2.8 mm long; trochanter 1.0 mm long; femur 5.80 mm long; tibia 7.0 mm long, hind tibial spurs present;tarsus 4.5 mm long, five segmented; claw 0.60 mm long, curved, dark brown.

Abdomen

18 mm long, 7 mm broad, dorsally brownish, densely covered with brownish scales on dorsal, ventral and lateral sides. Scales 0.18 mm long and 0.16 mm broad.

Genitalia (Figure 51)

1. 9.80 mm long, 5.80 mm broad, Uncus bifid, sparsely be set with setae on the dorsal side, apex nothched, down curved, hook like median dorsal process, chitinised, ending into teeth on each side.
2. Vinculum very short, saccus short and bulbus.
3. Harpes hinged to the lateral sides of vinculum, trilobed, first lobe muscular and flexible bearing numerous setae as well as strong bristles; dorsolateral lobe chitinous and bears two spines distally.
4. Tegumen broder in the middle, narrow at the both ends, apically the end produced into a flattened process, latter broadened at its end.
5. Anellus strong, chitinised, circular, anal opening chitinised into an almost quadrate plate.

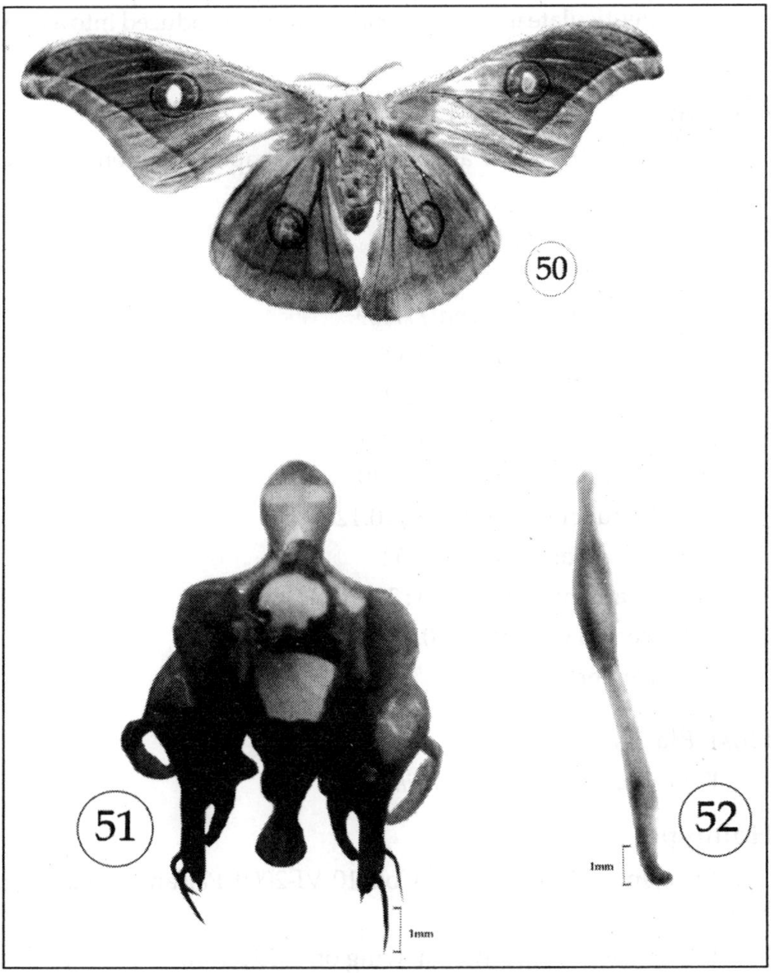

Plate 13

Figure 50: ***A. mylitta badami*–Adult male;**
Figure 51: ***A. mylitta badami*–Genitàlia;**
Figure 52: ***A.mylitta badami*–Adeagus**

6. Adeagus (Figure 52) 10.20 mm long, 0.85mm broad, narrow, basal part narrow than apical part, latter denticulate in its distal end, ventrally produced into a bow shaped plate.

Cocoon

Light grey in colour at surface, oval shaped, filament texture golden yellow.

1. Cocoon weight (g) : 7.80
2. Shell weight (g) : 0.60
3. Shell ratio (per cent) : 7.69
4. Cocoon length (cm) : 3.6
5. Cocoon width (cm) : 2.4
6. Peduncle length (cm) : 2.4
7. Peduncle width (mm) : 20
8. Peduncle weight (mg) : 0.12
9. Ring diameter (mm) : 11
10. Filament length (mt) : 310.15
11. Reeled weight (g) : 0.32
12. Denier : 9.28

Host Plants

Ber (Z. *jujuba*), Arjun (*T. arjuna*).

Holotype

Male, India, Maharashtra, Coll. 10-VI-2009, Palsambe –ramling, M.S,

Kavane. R.P., leg, antenna, wing on card sheet, body pinned in insect box, labeled as above.

Paratype

12 males: 2 females, sex ratio (m:f) 6:1, coll, from May to December, same data as above.

Etymology

The sub sp. name *badami* refers to its food plant badam.

Distributional Record

Western Maharashtra, Palsambe ramling 08-VII-2008; 1♀,1♂, Panhala 12-IV-2007; 1♂, Radhangari 28-VII-2009; 1♀,1♂, Patan 12-IV-2008; 1♂, Anuskara 02-IV-2006; 1♂, Atigre 2-IV-2007; 2♂, Dehu-alandi 28-IV-2007; 1♂, Kokrood 02-IV-2008; 2♂, Malakapur 12-IV-2007; 1♂, Amba 12-IV-2007; 1♂.

Remarks

Review of literature indicates that the present form runs close to *Antheraea mylitta* ecorace Sarihan by having following characters,

1) Cocoon weight 2) Shell weight 3) Silk ratio 4) Denier

However, it differs from above ecorace by having following characters.

1. Cocoon weight – 7.80 gm.
2. Shell weight – 0.60 gm.
3. Silk ratio – 7.69
4. Denier- 9.28
5. Morphological characters –
   (*a*) Hook like fore wing curvature
   (*b*) Medial line concave and strong
   (*c*) Hyaline area oval shape
   (*d*) Ocellus colour and radius
   (*e*) Genitalia shape and size

This sub species runs close to *A. mylitta sahydricus* by having two spines on dorsolateral lobe, fore wing acute or some what pointed, hyaline area not circular. However, it differs from *A.m. sahydricus* by following characters.

1. Saccus broder and shorter than *sahydricus*
2. Inner third lobe not straight, bent turn outwards
3. Adeagus faint and straight
4. Forewing curvature
5. Hyaline area
6. Adult body colour reddish yellow
7. Inner spine blunt

8. Adeagus long tapering anteriorly, swollen mid anteriorly.
9. Flagellar formula – 1 L/W = 0.33, 15L/W= 1.50, T L/W= 1.25, A = 1.02

ANTHERAEA MYLITTA SADHYCRICUS SUB SP.NOV

Male (Figure 53)

33 mm long, 9 mm broad, reddish brown dorsally, brown ventrally; antenna light brown,bipectinate, 17.5 mm long, 7 mm broad, 37 segmented; fore leg 21.8 mm long, wing expanse 136 mm.

Head

3 mm long, 5 mm broad, brown dorsally, ventrally grey; eyes rounded, large, light brown in colour, ocular distance 15 mm; proboscis vestigial; antenna 17.5 mm long, 7 mm broad, 37 segmented, terminally bifurcated, yellowish brown, scape 1 mm long, pedicel 0.40 mm long, flagellum 15 mm long; labial palpi upturned, three segmented, brownish.

Flagellar Formula

1 L/W = 0.33, 15L/W= 1.50, T L/W= 1.25, A = 1.02

Thorax

11mm long, 8 mm broad, dark brownish dorsally, greyish ventrally, prothrox, meso and metathorax brown with brown scales, scales 20 mm long.

Fore Wing (Figures 56i, 57i, 58i)

70 mm long, 43 mm broad, area of the fore wing 1382 mm^2, costal region of wing ash grey in colour; postmedian line pink with a white line on its border; antemedian line dark brown and bordered on the inside with a white line; oblique line brown with a indistinct faint white inner border; ventrally, pink, grey scales; antemedial line indistinct; medial line indistinct. Post medial line indistinct, margin in both wings brown. Ocellus (Figure 58i) 68 mm^2 with a hyaline area 35 mm^2, median cross vein concave and not touching with A1 nor A2, anterior outer line pink, anterior inner line white, anterior half dove grey, posterior outer line black. Posterior inner line light yellow, posterior half dove grey. Hyaline area circular;

wing scales conical with eight to nine spines of different lengths. Anterio lateral edge blunt not pointed, fore wing outer curvature (Figure 57i) not 'S' shaped. Wing expanse – 136 mm.

Hind Wing

40 mm long, 36 mm broad, area of hind wing 882 mm^2; area of ocellus 62 mm^2, area of hyaline spot 6 mm^2; hyaline area circular shaped; antemedian line is dark brown; oblique line absent.

Fore Leg

21.8 mm long, 1.6 mm broad, brown coloured; coxa 2.8 mm long; trochanter 1.0 mm long; femur 5.80 mm long; tibia 7.0 mm long, hind tibial spurs present;tarsus 4.5 mm long, five segmented; claw 0.60 mm long, curved, dark brown.

Abdomen

19 mm long, 9 mm broad, dorsally brownish, densely covered with brownish scales on dorsal, ventral and lateral sides, scales 0.18 mm long and 0.15mm broad.

Genitalia (Figure 54)

1. 9.30 mm long, 5.10 mm broad, Uncus bifid, sparsely be set with setae on the dorsal side, apex nothched, down curved, hook like median dorsal process, chitinised, ending into two pointed teeth on each side.
2. Vinculum very short, saccus long and rounded at its end.
3. Harpes hinged to the lateral sides of vinculum, trilobed, first two lobes are muscular and flexible bearing numerous hairs; dorsolateral lobe chitinous and bears two spines at distal extremity.
4. Tegumen broder in the middle, narrow at the both ends, apically the end produced into a flattened process, latter broadened at its end.
5. Anellus very strong, chitinised, circular, anal opening chitinised into an almost quadrate plate.
6. Adeagus (Figure 55) 9.70 mm long, 0.85mm broad, narrow, basal part shorter than apical part, latter denticulate in its distal on third part, ventrally produced into a bow shaped plate.

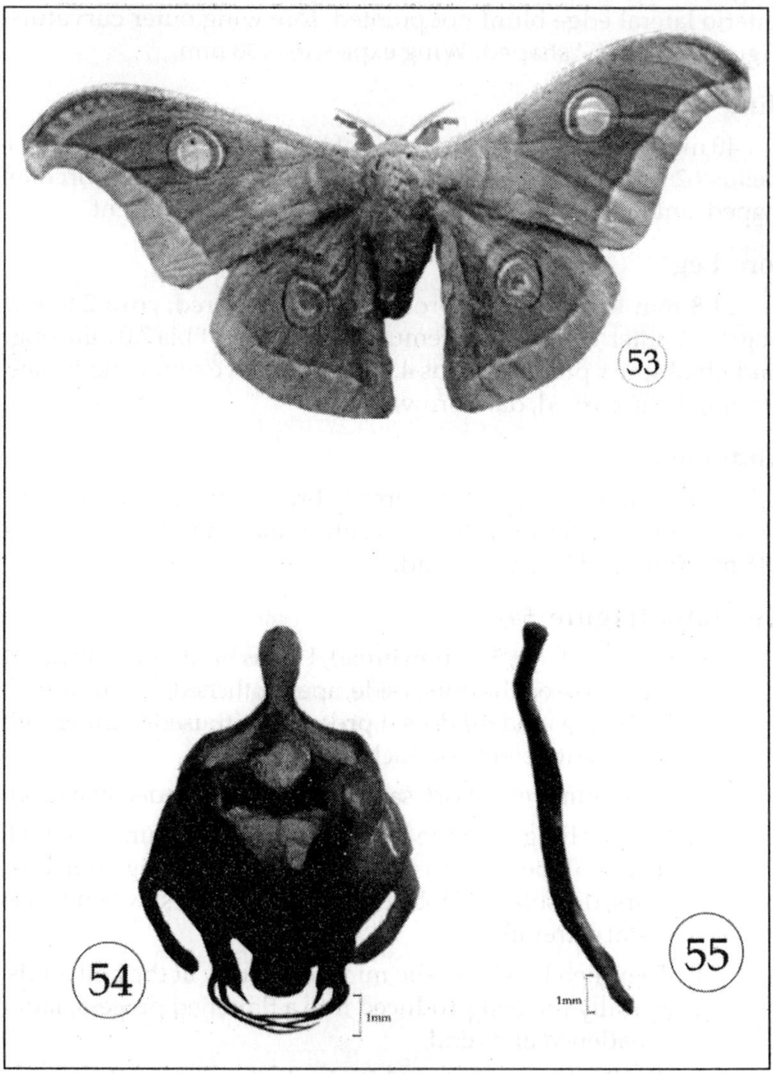

Plate 14

Figure 53: *A. mylitta sahydricus*–Adult male;
Figure 54: *A. mylitta sahydricus*–Gemialia;
Figure 55: *A. mylitta sahydricus*– Adeagus

Cocoon

Bright yellow in colour at surface, oval shaped, filament texture golden yellow.

1. Cocoon weight (g) : 7.65
2. Shell weight (g) : 0.90
3. Shell ratio (per cent) : 11.76
4. Cocoon length (cm) : 3.6
5. Cocoon width (cm) : 2.1
6. Peduncle length (cm) : 3.2
7. Peduncle width (mm) : 15
8. Peduncle weight (mg) : 0.09
9. Ring diameter (mm) : 8
10. Filament length (mt) : 311.20
11. Reeled weight (g) : 0.31
12. Denier : 8.96

Host Plant

Ber (*Z. jujuba*), Arjun (*T. arjuna*),

Asan (*T. tomentosa*).

Holotype

Male, India, Maharashtra, Coll. 18-VII-2008, Hatkanangle (ramling),

Kavane. R.P., leg, antenna, wing on card sheet, body pinned in insect box, labeled as above.

Paratype

12 males: 06 females, sex ratio (m:f) 2:1, coll, from May to December, same data as above.

Etymology

The sub sp. name *sahydricus* refers to the found in Sahyadri Ghat.

Distributional Record

Western Maharashtra, Hatkanangle ramling 18-VII-2008; 1♀,2♂, Panhala 12-IV-2008; 1♂, Radhangari 28-VII-2007; 1♀,1♂,

Patan 18-V-2007; 1♀,1♂, Anuskara 12-IV-2007; 1♀,Atigre 2-VI-2007; 1♀,1♂, Saswad 28-VIII-2008; 1♂, Kokrood 12-V-2009; 2♂, Malakapur 12-VI-2007; 1♀,2♂, Amba 12-VII-2007; 1♂.

Remarks

Review of literature indicates that the present form runs close to *Antheraea mylitta* ecorace Nowgaon by having following characters,

1) Cocoon weight 2) Shell weight 3) Silk ratio 4) Denier

However, it differs from above ecorace by having following characters.

1. Cocoon weight – 7.65 gm.
2. Shell weight – 0.90 gm.
3. Silk ratio – 11.76
4. Denier- 8.96
5. Morphological characters –
   (*a*) Hook like fore wing curvature
   (*b*) Medial line convex and strong
   (*c*) Hyaline area circular
   (*d*) Ocellus colour and radius
   (*e*) Genitalia shape and size

The sub species *A. mylitta sahydricus* runs close to *A.mylitta badami by* having two spines on dorsolateral lobe and general appearance of genitalia. However, it differs from above sub species by following characters.

1. Saccus very narrow
2. Dorsolateral spine very long
3. Harper broad, pointed, long
4. Adeagus dark and blackish
5. Forewing curvature
6. Hyaline area
7. Flagellar formula – 1 L/W = 0.33, 15L/W= 1.50, T L/W= 1.25, A = 1.02

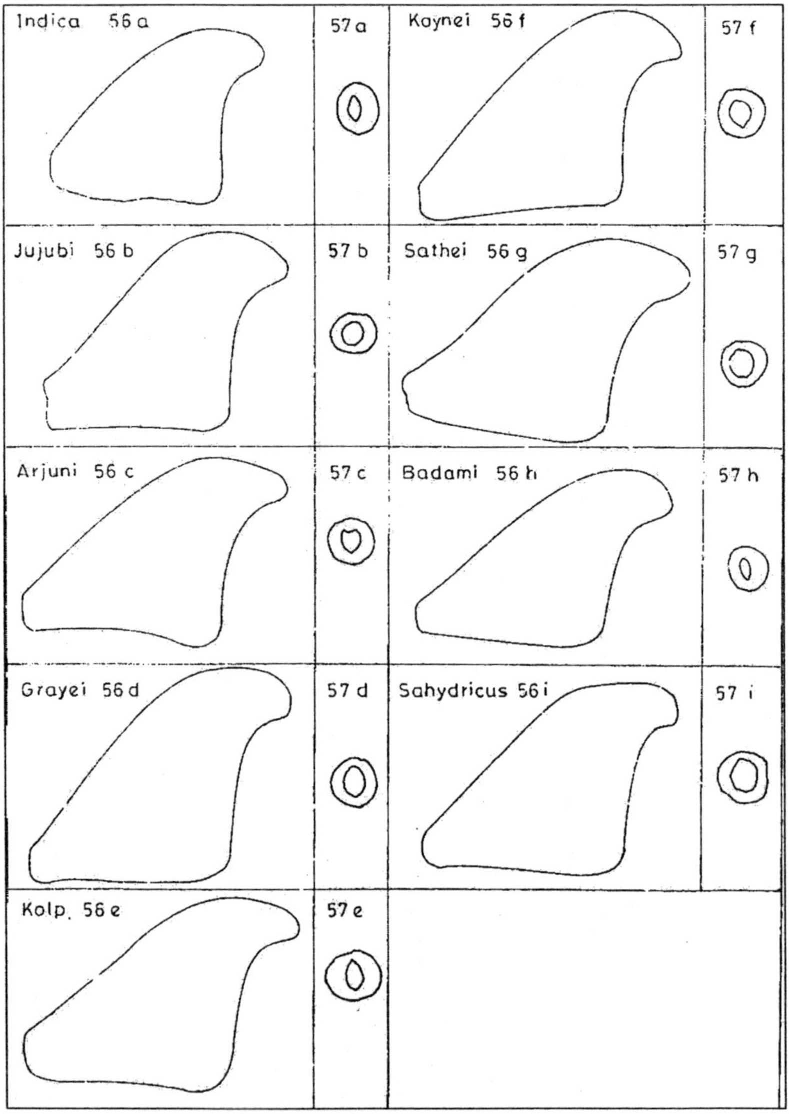

Figure56: a–i: Fore Wing of Wild Silkmoth
Figure57: a–i: Fore Wing Ocelli of Wild Silkmoth

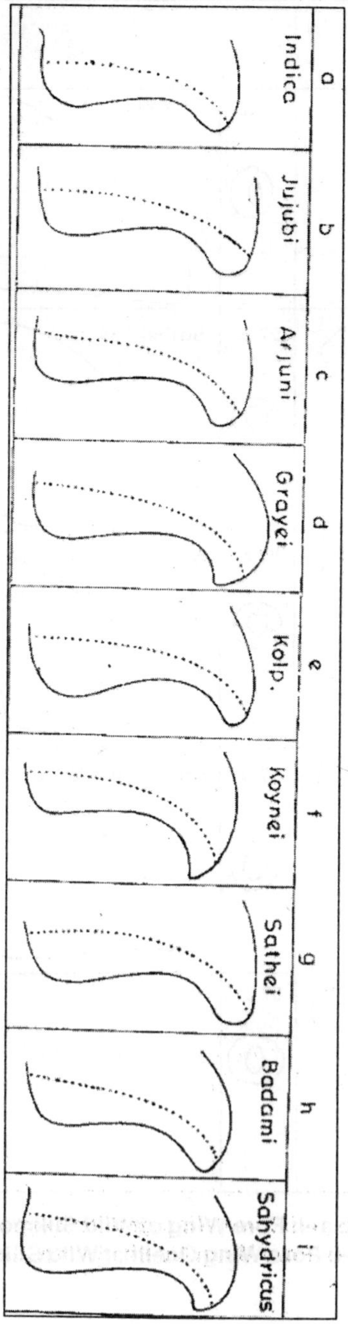

Figure58: a–i: Fore Wing Curvatures of Wild Silkmoth

Chapter 5

Biology of Wild Silkmoths

Silk, "the queen of textiles" is the natural fiber, which is secreted by silkworm, *i.e.* mulberry and non-mulberry silkworms (eri, muga, tasar and oak tasar) belonging to families Bombycidace and Saturnidace (Order–Lepidoptera, class-Insecta) respectively. Silk is proteineous in nature which is made up of fibrin ($C_{30}H_{46}N_{10}O_{11}$), fiber and sericin ($C_{30}H_{40}N_{10}O_{16}$). The filament is coated with gummy material. The silkworm rearing is called as sericulture which derived from the word "Su" (si) meaning silk is traditionally associated with the Socio-Economic life of many countries and used for exquisite textile and royal dresses since time immemorial (Chandra, 1997).

World wide 58 countries are engaged in silk production, where China with approximately 63 per cent of world total production is ranked top and India is second (approximately 16 per cent). China is known as origin country of silk which spreaded in 12th century B.C. to the other parts of the globe (Zhuang Da huan et al. 1994). India enjoy with all the five kinds of silk *i.e.* Mulberry, Eri, Muga, Tasar and Oak tasar, of these tasar named "Vanya Silk" is mostly practiced in tribal areas of tropical belt in India. Approximately 1.25 lakh tribal families are dependent on tasar for their lively hood and employment (Sathyanarayan and Rao, 2004). No doubt in recent year's industry emerged with a big role in improving Socio-economic status and generating employment opportunities.

Presently annual silk production in India is approximately 16319 MT, where tasar silk contribute 315 MT but, home need to silk (27700 MT) is more than our production (16319MT). So the enhancement in silk production is the time need but, there are set backs in the industry, like silkworm diseases, food quality, hybrid vigour, quarantine approaches, weather conditions and natural calamities etc. In addition, a need of skilled manpower always become the common factor.

In the recent years it has been noticed that HRD and management programmes are playing key role in enhancing productivity and quality improvement in various sectors by improving fitness and skill power of the workers *i.e.* officers, staff and labourers but, it is very complex process. It requires changing the orientation of managers, unions and workers development. Nadler (1984) stated need of training to increase job performance and growth. Pareck and Raw (1975) outlined 14 principles and processess of implementation in this line. Hence, the BTSSO Bilaspur an organisation of Central Silk Board (Govt. of India) also imparted a training programme for the sericulture officials in tasar sector.

North-eastern region has good potential for the growth of sericulture. Eri and muga silkworms still exist in the wild state in the forests of these regions (Thangavelu *et al.*, 1988). The Kotkori muga worms (*A. atlas*) feed on an unusual variety of food plant, The sole heather plant *Moyna laxiflora*, family Rubiaceae is distributed in the forests, on the road side and along the bank of many rivers and streams (Islam, 1990). The cocoon produced by this silkworm is some what similar to eri cocoon in nature and can be spun easily and other commercially important cut cocoons by using takli. The colour of this silk yarn resembles tasar spun silk. Sericulturists in this region have not cultured this variety. Little is known about the biology or product of this silkworm, Studies on the life cycle and proper maintenance of this silkworm may help the sericulture industry in the long run.

Saturniids, the wild silkmoths are having commercial importance in the silk production and their speciation is diversified worldwide. The wild Saturniid moon moth *A. selene* is geographically distributed and tropical, moist deciduous forest (Lefroy, 1909). Earlier, the moon moths were recorded from Mussorie and Sikkim in India (Cotes, 1891-1893), Afghanistan to Borneo (Jordan, 1911), Hong Kong

(Potter, 1941) and China, Japan and Sri Lanka (Essig, 1941). However, for the first time, their distribution in abundance was recorded from the forest of Bhandara, Maharashtra, Central India (Latitude 20.39°N–21.38° N: Longitude 79.29° E–80.42° E and Altitude 150-803 mts AMSL). Review of literature indicates that little is known on the life history and biology of wild sericinogenous insects in relation to their food plants. Hence, a detailed study was undertaken on the biology of *A. selene* on its primary host plants *T. arjuna* and *T. tomentosa; A. atlas* on *Ficus carica* and *A. mylitta* var. *kolhapurensis* on *T. catappa.* In past, Cotes (1891-93), Potter (1941), Waldbauer (1968), Jolly *et al.* (1968), Villiard (1969), Jolly *et al.* (1979), Barlow (1982), Nassig and Peigler (1984), Thangavelu and Sahu (1986), Peigler (1989), Murphy (1990), Akai *et al.* (1991), Siush and Thangavelu (1991),Thangavelu (1992, 1993, 1999),Jayaprakash *et al.* (1993), Rajadurai *et al.* (1998), Saikia *et al.* (1998), Babu and Rao (1998), Rath (1998), Baruah *et al.* (2000), Sathe and Jadhav(2001), Shamitha(2007) and Kavane and Sathe (2008) etc. worked on biology of silkmoths from India.

A. mylitta var. *kolhapuriensis* is reported from Kolhapur region which feed on leaves of arjun, ain, badam and ber, has been selected for biological studied, in Kolhapur, Sangli, Satara, Pune region. The above species found more potential for biological studies.

Materials and Methods

a) Biology of *Attacus atlas* (Linnaeus)

The cocoons of *A. atlas* were collected from a host plant of a Kolhapur forest region during the first week of July and the adults were allowed to emerge under laboratory conditions at 27 ± 2°C temperature and 75 ± 5 per cent RH (Gohain and Baruah, 1983) inside a moth cage made up of nylon. Female moths emerging from the cocoons were tied on along with their own cocoon and kept in the cage with male moth. Eggs collected from floor, walls of cage at room temperature 30 ± 2°C and RH of 80 per cent. Branches of Angier (*Ficus carica*) containing luxuriant leaves were cut under water and kept in conical flasks filled with water. The newly hatched larvae were released on the branches with the help of camlin brush and feather. Branches were changed from time to time and the rearing area was cleaned everyday. After maturity, the larvae were transferred to both dry and fresh leaf branches for spinning cocoons. Cocoons were harvested after 6 – 8 days of pupation.

b) Biology of *Actias selene* (Hubner)

Cocoons and female moths (both fertile and unfertile adults) were collected from Kolhapur forest region consisting of food plants, Arjun (*T. arjuna*) *and* Assan (*T. tomentosa*) and cultured in wooden cage (36 cm H × 36 cm L × 32 cm W) covered with closely perforated wire mesh. The cultural operations were conducted in the laboratory condition (25 – 30^{0}c, 70 – 80 per cent RH, 18 hr photoperiod).The adult female moths were confined to the earthen cups for egg laying. Grainage room was maintained at the optimum temperature range of 25-30°C and 70-80 per cent R.H. The newly hatched larvae were released on the branches with the help of camlin brush and feather. Branches were changed by 12 hr frequency and the rearing area was cleaned everyday. The data were recorded for the study of biology which includes the parameters such as adult longevity and larval life during the study period.

c) Biology of *A. mylitta* sub sp.nov *kolhapuriensis*

The cocoons of *A. mylitta* were collected from host plants,Arjun (*T. arjuna*), Assan (*T. tomentosa*) and *Ber* (*Z. jujuba*) from Kolhapur, Sangli, Satara and Pune region during the first week of April to June and the adults were allowed to emerge under laboratory conditions at 27 ± 2°C temperature and 75 ± 5 per cent RH inside a moth cage made of nylon. Eggs were collected from floor, walls of cage at room temperature of 27 ± 2°C and RH of 75 ± 5 per cent.

Newly hatched larvae were allowed to crawl on the leaves. Left over larvae were mounted on the leaves with the help of soft camel hairbrush. 4-5 such leaves with mounted larvae were placed on the wooden rods of the plastic box. The leaves were placed in such a way that the maximum portion of edge of each leaf was available to the larvae for feeding. The box was covered with its lid to prevent escape of the larvae. Next day the larvae were transferred to new box containing fresh leaf diet. The moulting larvae transferred along with their support leaves. Left over larvae were transferred with the help of a soft brush. The old leaves were removed from the boxes at 12 hr interval when the larvae crawled over a new leaves. The used boxes were then washed, disinfected and dried for re-use.

Rearing tray was used for rearing of 3rd, 4th and 5th instar larvae. 8 to 9 small twigs each having 12 to 15 leaves were used as leaf diet. Larvae were transferred to a new tray along with the remaining portion of diet leaf to avoid touching with hand. When they crawled to new leaves, the portions of old leaves were removed. Moulting larvae were transferred along with the portion for their attachment.

The full grown 5th instars wandering larvae were sorted and transferred to a new tray fitted with wooden rods and some twigs that provided them opportunity to form ring and peduncle. The data was recorded for the study of biology, which includes larval duration, adult longevity and fecundity. In all above species of silkworm sufficient number of silkworm (100) was used for confirming results. Morphological observations on immature and adults of silkworms species were made with the help of compound microscope and binocular.

Results

1) Biology of *Attacus atlas* L. (Figures 59 to 63)

Eggs (Figure 61)

The oval dorsoventrally compressed egg has a hard chitinised shell, composed of hexagonal cells. Egg is about 3.04 mm in length and 2.5 mm in breadth, weighing about 0.012 g. The newly laid eggs look creamy white but become light brown later. Eggs are laid scattered and have gummy substances due to which they get stuck to each other. Egg stage lasts for 10 days.

Larval Stage (Figure 62)

The first instar larva is 1.7 cm in length, 0.3 cm in breadth and 0.020 g in weight. Body colour is white with black inter segmental region. There are six tubercles in each body segment from 1st to 10th and five on 11th and four tubercles on the remaining 12th and 13th body segments. Each tubercle bears some hair like setae.

The late stage of 1st instar larva is quite different from the early stage. During late stage, it attains a length of 2 cm, breadth of 0.34 cm and weight 0.112 g. The early and late stage of this instar lasts for 1.5days and 3.5 days respectively. During late stage, the head of the tubercle becomes blunt and knob shaped. This instar lasts for 5 days.

Second instar larva is dorsally whitish in colour and some what orange-red on the lateral side of the body. Crystalline powdery substance was found on all the tubercles. The dorsal tubercles are bluish in colour. The head becomes light brown in colour. Second instar larva measures about 2.61cm in length, 0.4 cm in breadth and 0.785 g in weight. This stage lasts for 4.5 days.

Third instar larva is with five additional pairs of ventrolateral tubercles on each side of the first five body segments. The reddish colour get disappeared. The first three rows of dorsal tubercles gradually shortened. The clasper shows triangular shaped red ring. The larva measures 3.2 cm, 0.8 cm and 2.780 g in body length, width and weight respectively. This instar lasts for 4.0 days.

Fourth instar larva measures 5.52 cm in length, 1.3 cm in breadth and 5.58 g in weight. The dorsal tubercles of first three segments completely disappeared. All the lateral tubercles became black. Nine pairs of spiracles are seen. This stage lasts for 6.0 days.

Fifth instar larva (Figure 62) is 11.14 cm, 2.4 cm and 22.6 g in length, width, body weight respectively. Body colour becomes dark green and the lateral tubercle colour turns blue at the base and black at the tips. Dorsal tubercles become blue in colour. The thoracic legs are conical shaped with sharp distal claws. The 6th to 9th abdominal segments, each bears a pair of abdominal legs which are fleshy and flat. Terminally, they form a disc with a series of hooks inwardly curved and arranged in a semi-circle. However, a pair of white spots appeared on the ventral side of the 11th abdominal segment in the female larva but, a only single spot is visible in male. The dorsal tubercles project backwards and lateral tubercles project forward. The instar fifth lasts for 7 days.

Pupa and Cocoon (Figure 63)

The mature larva construct its cocoon on fresh leaves and suspends it from the twig with the help of a long stalk. It spins its silk fibre around its body with the help of spinneret and tubercles. The pupa is brown coloured and 4.4 cm, 1.4 cm and 7.6 g in length, width and body weight respectively. This stage lasts for 28 days. The colour of cocoon is greyish brown. *A. atlas* silkworm undergoes pupation in an open type silk cage.

Moth (Figure 60, 61)

Moth emergence takes place in the early morning and just after emergence it clings to its own cocoon and remains there for 8-10 hours till its wings are fully stretched. The males are more active and couples with the females at dusk which lasts for 12-24 hours. The female (Figure 60) lays 134 eggs which are scattered on the sides of cage. The male moth survive for 2-3 days and the female for 4-6 days after copulation and of egg laying respectively. The red brown moth has a wing span of 26 cm in male and 28 cm in female. The basal area of the forewings is brown and red brown edged with red, pale and black lines. Medial area is red brown. At the end of cell with a black edge a large hyaline spot is present. Apical area has yellow to pink shade. A yellow brown marginal band with a highly waved black line is present on the fore and hind wings.

2) Biology of *Actias selene* Hubner (Figures 64 to 68)

Eggs (Figure 65)

The eggs are oval and dorsoventrally compressed which has a hard chitinised shell. Egg is about 3.04 mm in length and 2.8 mm in breadth. The newly laid eggs are grey in colour but later turn dark. Eggs are laid scattered and attached to each other by gummy substances. Egg stage lasts for 8 days.

Larval Stages (Figures 66, 67)

The newly hatched instar larva is 7 mm in length, 1mm in breadth and 15 mg in weight. Body colour is orange brown with two black bands. Tubercles bearing hair like setae.

The late stage of 1st instar larva (Figure 66) looks quite different from the early stage. Late stage larva measures 1.8 cm, 1mm and 0.105 g in body length, breadth and weight respectively. Body colour is orange brown with two black bands. This stage lasts for 4.5 days.

Second instar larva body colour is orange brown, black bands disappeared. The head becomes light brown in colour. Second instar larva measures about 2.12 cm in length, 2 mm in breadth and 0.453 g weight. This stage lasts for 5 days.

Third instar larva is green with orange tubercles. The first three rows of dorsal tubercles gradually become shorter. Larva reaches a length of 2.7 cm, breadth 2 mm and weight 1.780 g. This instar lasts for 6 days.

Fourth instar larva measures 3.91 cm in length, 3 mm in breadth and weight 5.58 g. Larva is bright green in colour. This stage lasts for 6 days.

Fifth instar larva (Figure 67) is 7.56 cm in length, 1.8 cm breadth and weight 18.6 g. Body colour becomes dark green and the lateral tubercle colour turns orange. Dorsal tubercles appear yellow in colour. The thoracic legs are conical shaped and carry sharp distal claws. This stage lasts for 10 days

Pupa and Cocoon (Figure 68)

The mature larva builds its cocoon on fresh leaves and suspends it from the twig with the help of a long stalk. The mature larva spins its silk fibre around its body with the help of spinneret and tubercles. The pupa is brown in colour, length is 3.4 cm and breadth 1.7cm and weight about 8.6 g. This stage lasts for 20 days. The cocoon colour is whitish grey. It measures 4.2cm length and 2.2 cm width. Cocoon weight 8.65 g. The life cycle from egg to adult was completed with 49.5 days in *A.selene* on host plant *T. arjuna, T. tomentosa* and *T. catappa.* Three generations have been completed in the laboratory.

Moth (Figure 64)

As like *A. atlas* emergence of moth takes place in the early morning and just after emergence, the males become more active and couples with the females at dusk which lasts for 16 hours. The copulated female lays about 105 eggs. The male moth dies after 3-4 days copulation and the female after 3-5 days of egg laying. The pista coloured moth has a wing span of 5.4 cm in male and 5.7 cm in female. Frontal tuft is white. Antennae are pale brown with whitish antennal base, fore wings are about pale green and legs are pinkish on the upper side and pale yellowish on underside.

3) Biology of *Antheraea mylitta* sub sp. nov *kolhapuriensis* (Figures 69 to 74)

Eggs (Figure 70)

The egg is comparatively large, oval, dorsoventrally flattened bilaterally symmetrical along the antero-posterior axis and with hard chitinised shell. It measures about 3.04 mm in length and 2.5 mm in breadth and weighing about 9.6 mg. The newly laid eggs are dark brown, after washing it becomes white, light yellow or creamy. The egg shows two dark brown lines running parallel to each other. Incubation period is 8 days.

Plate 15: Life Cycle of *Attacus atlas* L.

Figure 59: Mating (a–male; b–female); Figure 60: Female Silkmoth; Figure 61: Eggs; Figure 62: 5th Instar Larva; Figure 63: Cocoons

Plate 16: Life Cycle of *Actias selene* Hubner

Figure 64: Female Moth; Figure 65: Eggs; Figure 66: 1st Instar Larvae; Figure 67: 5th Instar Larva; Figure 68: Cocoon

Plate 17: Life Cycle of *Antheraea mylitta kolhapurensis*

Figure 69: Female Silkmoth; Figure 70: Eggs and 1st Instar Larvae; Figure 71: 3rd Instar Larvae; Figure 72: 4th Instar Larva; Figure 73: 5th Instar Larva and Cocoons; Figure 74: Pupa

Larval Stage (Figures 70, 71, 72, 73)

The newly hatched larva is elongated, cylindrical being flattened on the ventral surface, robust, hairy and clothed with numerous tubercles. Body of newly hatched larva (Figure 70) shows dull brownish –yellow body colouration with a blackish head. Head capsule is glossy black with bristles. Clypeus is white and labrum brown. Diameter of head capsule is 1.52 mm. Early stage larva measures about 0.70 cm in length, 0.30 cm in breadth and 82mg in weight. Late first instar has thoracic legs black with reddish claw.

The appearance of the late stage of 1st instar larva became quite different from the early stage. In the early first instar larva there is a black mid-dorsal line extending from the first to seventh abdominal segment and body colour is greenish. Colour of tubercles is black. The late first instar measures 1.31cm in length, 0.22cm in breadth and 322mg in weight. The head capsule of this stage is brown coloured with white bristles and with diameter of head capsule 1.52 mm. Oval black spot appeared on prothoracic hood dorsally. This stage lasts for 3 days.

Second instar larva is with pale brown head capsule. Eyes are black with white bristles. Head capsule diameter is 2.30 mm. Larva measures about 2.41cm in length and 0.31cm in breadth and its weight is 341 mg. Early second instar prothoracic hood bears M shape and later V shape with two black dots. Tubercle colour is orange red. Early second stage lasts for 3 days.

Third instar larva (Figure 71) is 3.65 cm long and 0.52 cm broad, and weighed 1.28g. Dorsally on prothoracic hood there is no any mark. Diameter of head capsule 3.40 mm, chocolate coloured with distinct mid cranial inflection, adfrontal sulcus frons are sienna and the palpi are greyish. Tubercles are violet in colour. This stage lasts for 6 days.

Fourth instar larva measures 7.82 cm in length, 1.41 cm in breadth and 6.58 g in weight. Dorsally prothoracic hood shows two semilunar red marking corresponding to the epicornial plates. Diameter of head capsule is 5.30 mm and head capsule is chocolate coloured. Tubercles are violet in colour. This stage lasts for 6 Days.

Fifth instar fully grown larva (Figure 72) is 12.14 cm in long, 1.94 cm wide and 19.95 g in weight. Body colour is dark green and the lateral tubercle colour turns blue at the base and dorsally prothoracic hood appear two semilunar red marking. Diameter of head capsule is 6.80 mm and head capsule is chocolate coloured. Fully grown fifth instar larva starts spinning cocoon around it self. Fifth stage larva lasts for 8 days.

Pupa (Figure 74) and Cocoon (Figure 73)

The pupa is brown in colour. Its length is 4.1 cm and breadth 1.5 cm and weight 6.6 g. Larvae were sorted out and transferred to a new tray fitted with old twigs that provide them opportunity to form ring and peduncle. The matured larva spins its silk fibre around its body with the help of spinneret and tubercles. This stage lasts for 23 days. The cocoon colour of silkworm is whitish grey, length of cocoon is 4.8 cm and width 2.9 cm and cocoon weight is 11.35 g.

Moth (Figure 68)

In the early morning moth emergence takes place. The males became more active and coupled with the females at dusk which lasts for 18 hours. The female lays about 185 eggs in laying box. The male moth dies after 4-5 days after copulation and the female after 3-4 days of egg laying. Moth has a wing span of 12 cm and 14 cm in female.

Discussion

Saikia *et al.* (1998) studied the life cycle of *A. atlas* by providing main food plant *Meyna laxiflora* under which the incubation period of eggs was 10 days, the larval period was 28 days and the pupal duration was 28 days. The adult male survived for 2-3 days and female 4-6 days. Jolly *et al.* (1979) reported grayish brown colour of the cocoons of *A. atlas* insect. While, Hampson (1892) reported tubercular arrangement in the larva of *A.atlas*.

Peigler (1989) reported over 100 plant species belonging to 90 genera in 48 families as host plants for *Attacus* spp. Villiard (1969) was of the opinion that greater success on the rearing of *Attacus* larvae particularly the later instars could be achieved by feeding them on a mixed diet of above said plant. Murphy (1990) was the first to mention the presence *of Attacus* in mangrove habitats, stating that *A. atlas* occurred once on *Avicennia alba* Bl. (Avicenniaceae) and simultaneously with many other trees. However, it occurred at low levels on *Bruguiera gymnorhiza* (L.) Lamk. (Rhizophoraceae). In the present study biology of *A. atlas* was studied on host plant *Ficus carica* and the silkworm completed its life cycle from egg to adult within 64 days.

Rajadurai *et al.* (1998) studied the life cycle of *A. selene* and reported that *A. selene* is distributed widely all along the mixed forests plants such as *T. arjuna, T. tomentosa* and *Z. mauritiana.* The incubation period of egg was 9 days, the larval period was 31days and the pupal duration was 18 days. The adult male survived for 3-4 days and female 4-6 days. The total period for completion of life cycle was 58 days. While, in the present study the life cycle from egg to adult was completed in about 49 – 50 days which is considerable shorter than the previously recorded.

Cotes (1891-93), Barlow (1982) studied the host plants of *A. selene. A. selene* feed on *Zanthoxylum acanthopodlum* D C., *Z. alatum* Roxb (Rutaceae), *Cedrela paniculata* (Meliaceae). *Coriana nepalensis* Wall. (Coriariaceae), wild cherry Prunus, Wild pear Pyrus (both Rosaceae), walnut (Juglandaceae) and other fruit trees in Northern India.

Nassig and Peigler (1984) stated that some members of Anacardiaceae are good host plants for *A.selene.* Host plants *Heptapleurum octophyllum* B. and H. (Araliaceae) were reported from Hong Kong (Potter, 1941). Chen-Shuren *et al.* (1997) observed in China that larvae of *A. selene ningopoana* Felder caused considerable damage to the plant, *Cornus officinalis* Sieh at Zucc by feeding on leaves. The moon moth (*A. selene*) pupa undergoes summer diapause. The moth exhibits trivoltine nature which is to synchronize with the meteorological condition of the region. In the present study only three generations have been completed.

Jolly *et al.* (1968) reported that tasar silkworm *A. mylitta* is polyphagous in nature feeding on a number of host plants. *T. arjuna* (Arjun), *T. tomentosa* (Asan) and *S. robusta* (Sal) are best among all the food plants and are considered as primary food plants for commercial rearing. They have since been adopted for large scale exploitation for tasar rearing in the country. Thangavelu (1992, 1993, 1999) discussed the need for conservation of wild sericigenous insects of India and also indoor rearing of tasar silkworm (*A.mylitta*) for domestication. He also highlighted the importance of systematic research on various aspects of non-mulberry silk.

Babu and Rao (1998) described Indoor chawki rearing of tropical tasar silkworm *A mylitta* in a specially designed Indoor Chawki Rearing Tray (ICRT) with hydrophonic branches as a new concept. A comparative study of indoor-cum-outdoor and field rearing of *A.mylitta* have been reported. There is a significant improvement in the Effective Rate of Rearing (ERR) when the worms were reared on ICRT with hydrophonic branches up to the 3rd instar under indoor conditions and later, on outdoor plants. Rath (1998) reported first of its kind in case of *A. mylitta* for the evaluation of food plants using growth indices parameters. Earlier attempts were merely based upon feeding trials for evaluation of the host plant for tropical tasar silk worms. The actual performance of an insect reduced below its physiological potential by food quality (Slansky and Scriber, 1984).

Reproduction in insects is very closely related to nutritional factor, the qualitative and quantitative aspects of that have an impact not only on fecundity but also on the rate of growth and development.The silkworm larvae are known to be continuous feeders, hence, the deficiency in the amount of food required to reach its full potential will be manifested in various ways and degrees (Waldbauer, 1968). According to Waldbauer (1968) the control plant *T. tomentosa* provides the necessary nutritional requirements for better larval growth, survival, pupal and adult weight, adult emergence and ovipositional potency for *A. mylitta*.

Kavane and Sathe (2008) reported rearing technique for tasar silkworm *A. mylitta*. The results indicated that the rearing success of *A. mylitta* on *T. catappa* under laboratory conditions (24 ± 1 °C, 65-70 per cent R. H. and 14 hr photoperiod) was 45 per cent. The cocoon quality was satisfactory. The silkworms were adopted in indoor rearing technique by preparing no peduncle which is normally spun by the worms in outdoor rearing is outstanding feature of the success of indoor rearing technique.

In past, attempts have been made by some workers (Chanda and Gupta, 2005; Sathe and Jadhav, 2001; Shamitha, 2007, Kavane and Sathe, 2008 etc.) to develop indoor rearing method for tasar silkworm *A. mylitta*. Institute of sericultural and an Entomological science, Japan has developed some artificial diet for tropical tasar silkworm containing Asan leaf powder, the principal food plant, and has achieved success to some extent (Akai *et al*., 1991).

Jayaprakash *et al.* (1993) got 20 per cent success in increasing effectiveness of rearing of *A. mylitta* by putting check on natural enemies of silk worm. In field condition, young age worms "Chawki worms"; are very susceptible for pest attack. Siush and Thangavelu (1991) studied the biocontrol potential of natural enemies of tasar silkworm in the field condition. Shamitha (2007) attempted total indoor rearing of tasar silkworm *A. mylitta*. Shamitha (2007) reported that tasar culture in a forest based industry best suited to the economy and social structure of developing countries like India. Biometrical measurements were taken in the present forms, considering the structure of head capsule, body length and width of larval stage. Head capsule, length and width of larval body were frequently chosen as characters which may help in determining the number of larval instars. The head capsule and larval length and breadth were age increase instar wise. The most sclerotic part like head capsule also increased in diameter as per the age.

In the present study, biology of 3 silkworms namely *A. atlas*, *A. selene* and *A. mylitta kolhapurensis* have been studied and result indicated that all 3 species of silkworm complete their life cycle in relatively short period than that of previously reported. Hence, the life cycle and biological studies of above silkworms will be helpful for developing standard rearing technique for silkworms and in understanding the biological peculiarities of the silkworms.

Chapter 6

Rearing Techniques for Wild Silkmoths

India is the second largest producer of the tasar silk in the world (Shankar Rao *et al.*, 2004). This silk is obtained from the tasar silkworm *Antheraea mylitta* Drury which prevails in central India distributed in moist semi deciduous forests of Jharkhand, Chhattisgarh, Orissa and the fringes of Andra Pradesh and Maharashtra. There are about 44 ecoraces of which 14 are exploited commercially. Survey, Protection, Conservation and exploitation of more and more ecoraces of tasar silkworms is essential for sustainable development of out country. Secondly, silk derived from tasar is economically viable which protect the man from certain diseases and bad effects of ultraviolet rays. Hence investigating standard technique for mass rearing of tasar silkworm *A. mylitta* and other silkworms like *A.atlas* and *A.selene* is challenge to Sericulturists and scientific community.

In past, Jolly (1972), Ahsan *et al.* (1974, 2000), Akai *et al.* (1990), Das and Nayak (1991), Lin yan *et al.* (1992), Jayaprakash (1995), Thangavelu *et al.* (1993), Ojha et al.,(1994), Pujar and savanurmath (1998),Srivastava *et al.* (1998), Patil (1998), Tiwari (1998), Mathur et al.,(1998), Nayak and Jaganth (1998),Alam *et al.* (1998), Sudhakar and Purushtam (1998), Rath (1998),Sinha U.S.S.P *et al.* (1998), Reddy *et al.* (1998), Purushtam Rao (1998), Madan *et al.* (1998), Rajadurai and Thangavelu (1998), Saikia and Handique (1998), Sathe and Jadhav (2001), Shimata (2007), Kavane and Sathe (2008), etc. attempted the work related to rearing of wild tasar silkworms.

Materials and Methods

For rearing of wild silkmoths *A. atlas, A. selene* and *A. mylitta kolhapurensis* the basic plan of methodology adopted is the same as maintained in previous chapter *i.e.* Biology of wild silkmoths. Other details of rearing of wild silkmoths of each species is given in materials and methods chapter III and Tables 2–4.

Cocoon Characterization

Following characters were taken into account for assessment of rearing potential of wild silkworm species.

(*i*) Single cocoon weight

The average cocoon weight in grams of 10 cocoons taken at random from each replicate on sixth day of spinning was considered as the cocoon weight.

(*ii*) Single Cocoon Shell Weight

The trait requirements, the total quantity of silk in a cocoon. The cocoon shell weight of 10 cocoon are once again used for calculating the average shell weight, which is expressed in grams.

(*iii*) Determination of the Size of the Cocoon

Length and width of the cocoon shell was determined with vernier calipers. The width of the cocoon was measured in different areas such of head, middle, tail regions and calculated by using formula:

Total length/width= Main scale reading + vernier coincidence × least count.

(*iv*) Determination of the Thickness of the Cocoon Shell

The thickness of the cocoon shell was determined with the help of the screwguage. The cocoon was cut into small pieces in different regions like head, tail, side-1(right side) and side-2 (left side) to measure its thickness and calculations were made by using the following formula:

Table 2: Requirements for Rearing One DFL for *A. atlas* Silkworm*

Instar of Worm	*Duration (Days)*	*Feeding Time per Day*	*Feeding Time Total*	*Feeding Dose Total (kg)*	*Leaf Proportion (g)*	*Leaf Number on Food Plant Twig*	*Leaf Size*	*No. of Boxes/ Cage*	*No. of Trays*	*Box/ Tray Cleaning Time*	*Duration of Shedding Cuticle (hr)*	*Humidity*	*Temp. °C*	*Bed Size sqft*
1st	5	1	5	1–1.5	40–50	2–3	whole	1	–	2	24	75	28–30	1sqft
2nd	4.5	2	9	2–2.5	50–60	3–4	whole	1	–	2	48	75	28–30	3sqft
3rd	4	2	8	3–4	50–60	Medium/ June	whole	–	1	3	72	80	28–30	6sqft
4th	6	2	12	4.5–5	65–75	Medium/ June	whole	–	1	Once every morning	72	80	28–30	10sqft
5th	7	2	18	4.5–5	75–85 June	Medium/	whole	–	1	Once every morning	–	80	28–30	16sqft

One dfl contain–50 eggs for *A.atlas* silkworm, Food plant used–Angier (*Ficus carica*)

Table 3: Requirements for Rearing One DFLs for *A. selene* Silkworm*

Instar of Worm	*Duration (Days)*	*Feeding Time per Day*	*Feeding Time Total*	*Feeding Dose Total (kg)*	*Leaf Proportion (g)*	*Leaf Number on Food Plant Twig*	*Leaf Size*	*No. of Boxes/ Cage*	*No. of Trays*	*Box/ Tray Cleaning Time*	*Duration of Shedding Cuticle (hr)*	*Humidity*	*Temp. °C*	*Bed Size sqft*
1st	4.5	1	3	1–1.5	40–50	2–3	whole	1	–	2	24	75	26–28	1sqft
2nd	5	2	6	2–2.5	50–60	3–4	whole	1	–	2	48	75	26–28	2sqft
3rd	6	2	14	3–4	50–60	Medium/ June	whole	–	1	3	72	80	26–28	6sqft
4th	6	2	14	5.5–6	65–75	Medium/ June	whole	–	1	Once every morning	72	75–80	26–28	12sqft
5th	10	2	22	6.5–7	75–85	Medium/ June	whole	–	1	Once every morning	–	75–80	26–28	18sqft

* One dfl contain–100 eggs for *A. selene* silkworm, Food plants used–*T. catappa, T. arjuna, T.tomentosa*

Table 4: Requirements for Rearing One DFLs for *A. mylitta kolhapurensis* Silkworm*

Instar of Worm	*Duration (Days)*	*Feeding Time per Day*	*Feeding Time Total*	*Feeding Dose Total (kg)*	*Leaf Proportion (g)*	*Leaf Number on Food Plant Twig*	*Leaf Size*	*No. of Boxes/ Cage*	*No. of Trays*	*Box/ Tray Cleaning Time*	*Duration of Shedding Cuticle (hr)*	*Humidity*	*Temp. °C*	*Bed Size sqft*
1st	3	1	3	1–1.5	100–150	3–4	whole	1	–	2	30	85	26–28	1sqft
2nd	3	2	6	4–4.5	150–200	Medium	whole	1	–	2	48	80	26–28	3sqft
3rd	6	2	12	8–8.5	200–300	Medium/ June	whole	–	1	3	72	80	26–28	6sqft
4th	6	3	18	10–14	300–400	Medium/ June	whole	–	2	Once every morning	96	75–85	26–28	12sqft
5th	8–9	3	27	23–25	300–400	Medium/ June	whole	–	2	Once every morning	–	75–85	26–28	18sqft

* One dfl contain–150 eggs for *A. mylitta* silkworm, Food plants used–*T. catappa, T. arjuna, T.tomentosa, Z. jujuba.*

Thickness of the cocoon = Pitch scale reading + Head scale reading × least count

(*v*) Shell Ratio

$$\frac{\text{Weight of shell}}{\text{Weight of cocoon}} \times 100$$

Fecundity

Fecundity was studied with the help of offspring's developed from single mated female.

Longevity

Longevity of adults in the laboratory was studied by providing no food to moths.

Measurements of Consumption and Utilization

The gravimetric method as described by Waldbauer (1968) was used to measure the utilization of food on wet weight basis by the larvae.

1. Weight of larvae, initial weight of leaves given as food, weight of leaves after consumption and weight of litter were measured every 24 hr at around 8 A.M.
2. The larvae weight gain during an instar was calculated by subtracting its weight at the beginning of a feeding period from its weight attained before settling into moult.
3. Weight of the leaf diet kept separately in a tray without larvae under similar condition was also recorded on next day to estimate the moisture loss or "natural loss" of food.
4. As the food consumption and egestion during moulting period is nil, it was not taken into account.
5. Litter were separated from the uneaten food and weighed.
6. All weights were taken in gm.
7. Food energy assimilation (FEA) = food consumed (F) – food egested (E)
8. Respiration (R) = Food consumed (F) – {Food egested (E) +Increase of larval weight (G)}.

Abbreviations Used in the Text

A: Weight of larvae
F: Food consumed
G: Weight gain
E: Excreta
R: Respiration
FEA: Food energy assimilation

Results

a) Rearing of *A. atlas*

The results tabulated in Table 5 indicate that the cocoon formation takes place within 35 to 43 days (average 40 days) cocoon weight, shell weight, length of shell, width of shell and shell thickness was 9.42g, 1.82g, 4.4cm, 1.4cm and 0.21mm respectively while, shell ratio was 19.32 per cent. The rearing success of *A.atlas* silkworms on Angier (*Ficus carica*) leaves was 2 per cent (average 1-3 per cent). Other food plants *T. arjuna, T.tomentosa* and *T. cattapa* were tried but silkworm refused as food under laboratory condition (Table 2). On an average a single mated female produced 25 offsprings with an average sex ratio (m:f) 1:0.66. The longevity of adult moths averaged 4 days in males and 6 days in females.

Table 5: Cocoon Characterization of *A. atlas*

Sl.No.	*Cocoon Wt (g)*	*Shell Wt (g)*	*Length of Shell (cm)*	*Width of shell (cm)*	*Shell Thickness (mm)*	*Shell Ratio (per cent)*
1.	9.42	1.82	4.4	1.4	0.21	19.32

b) Rearing of *A. selene*

Results are recorded in Table 6. The cocoon formation in *A.selene* was started from 45th day after egg hatching and stopped spinning on 48th day (average 45 days), cocoon weight, shell weight, length of shell, width of shell and shell thickness was 7.42g, 0.93g, 4.2cm, 2.2 cm and 0.18mm on the food plant *T. catappa*. It was 8.05g, 1.20g, 3.9cm, 2.1cm and 0.19mm on the food plant *T. arjuna*. While on *T.tomentosa* it was 8.35g, 1.41g, 3.8cm, 2.1cm and 0.21mm respectively.

The shell ratio of cocoons on above plants was 12.53 per cent, 14.40 per cent and 16.88 per cent respectively. The rearing success of *A. selene* silkworms on food plants *T. arjuna, T.tomentosa* and *T. cattapa* was 5 per cent (average 4-6 per cent), 5 per cent (average 4-6 per cent) and 2 per cent (average 1-3 per cent) respectively under laboratory conditions (Table 3).

Table 6: Cocoon Characterization of *A.selene*

Sl.No.	Food Plant	Cocoon Wt (g)	Shell Wt (g)	Length of Shell (cm)	Width of Shell (cm)	Shell Thickness (mm)	Shell Ratio (%)
1.	*T. catappa*	7.42	0.93	4.2	2.2	0.18	12.53
2.	*T. arjuna*	8.05	1.20	3.9	2.1	0.19	14.90
3.	*T.tomentosa*	8.35	1.41	3.8	2.1	0.21	16.88

c) Rearing of *A. mylitta Kolhapurensis* (Figures 75-98)

Results are recorded in Table 7. The cocoon formation in *A. mylitta kolhapurensis* was started from 38th day and stopped on 42nd day (average 38 days). Cocoon weight, shell weight, length of shell, width of shell and shell thickness was 10.42g, 1.09g, 4.1cm, 2.4cm, and 0.28mm on the food plant *T. catappa*; 11.32g, 1.32g, 3.9cm, 2.3cm and 0.32mm on the food plant *T. arjuna* and 11.42g, 1.56g, 3.8cm, 2.4cm, and 0.33mm on *T. tomentosa* respectively. The shell ratio was 10.46 per cent on *T. catappa*, 11.46 per cent on *T.arjuna* and 13.66 per cent on *T.tomentosa*. The rearing success of *A. mylitta kolhapurensis* on food plants *T.arjuna, T.tomentosa* and *T. cattapa* was 43 per cent (average 35-60 per cent), 40 per cent (average 35-60 per cent) and 45 per cent (average 35–60 per cent), respectively under laboratory conditions (Table 4).

Table 7: Cocoon Characterization of *A. mylitta kolhapurensis*

Sl.No.	Food Plant	Cocoon Wt (g)	Shell Wt (g)	Length of Shell (cm)	Width of Shell (cm)	Shell Thickness (mm)	Shell Ratio (%)
1	*T. catappa*	10.42	1.09	4.1	2.4	0.28	10.46
2	*T. arjuna*	11.32	1.32	3.9	2.3	0.32	11.66
3	*T.tomentosa*	11.42	1.56	3.8	2.4	0.33	13.66

Daily Food Consumption and Utilization per Larva of *A. atlas* reared on *F. carica* (Angier)

Food Consumption (F)

A larva of *A.atlas* fed on Angier consumed 0.0543g of leaf on first day, 0.0763 g on second day, 0.0869 g on third day, 0.1284 g on fourth day of feeding. Decline in food consumption was observed on the fifth day, which was the last day of the first instar. There was progressive increase in food consumption from sixth day 0.3707g to eighth day 0.6490 g, from tenth day 0.5105 g to fourteenth day 2.7208 g and fifteenth day 1.9023 g to eighteenth day 4.0555 g of feeding. From twenty first day 2.9980g to twenty sixth day 8.7554 g food consumption was irregular. Maximum food was consumed on twenty-sixth day 8.7554 g.

Mean Weight of Larva (A)

Weight of freshly hatched larva of *A.atlas* was 0.0083g. On subsequent days it increased progressively till twenty sixth days when the larva attained the maximum weight of 13.6351 g in fifth instar. Larval weight on day fifth, ninth, tenth, fifteenth, sixteenth, twenty-first, twenty second and twenty-sixth day was 0.0841 g, 0.4560 g, 0.4513 g, 2.8173 g, 3.1797 g, 7.3610 g, 8.4750 g and 13.6351 g respectively.

Weight Gain (G)

Initial weight gain in larva was calculated to be 0.0049g on second day. It increased daily till thirteenth day 0.4466 g and decreased on sixteenth day 0.3623 g which was the third day of the fourth instar. In fourth instar also the weight gain decreased on the third day only *i.e.* on sixteenth day 0.3623 g, Weight gain again increased from ninetieth day 0.7310 g. Afterwards decrease in weight gain was observed till twenty-sixth day. Negative weight gain was observed on twenty-fourth day. Maximum weight gain was calculated on twenty-second day 1.1130 g and minimum on twenty-fourth day –0.0233 g indicating negative scope for growth during this period.

Excreta (E)

Weight of litter on second day was found to be 0.0069 g. It progressively increased on subsequent days till thirteenth day 1.0190

Table 8: Daily Food Consumption and Utilization Per Larva of *A. atlas* reared on *F.carica* (Angier)

Unit: g.

Day of Feeding	*F*	*A*	*G*	*E*	*FEA*	*R*
1st	0.0543	0.0083	–	–	–	–
2nd	0.0763	0.0132	0.0049	0.0069	0.0694	0.0645
3rd	0.0869	0.0272	0.0140	0.0156	0.0713	0.0573
4th	0.1284	0.0619	0.0347	0.0760	0.0524	0.0177
5th	0.1265	0.0841	0.0232	0.1041	0.0224	–0.0008
First moult period of 1 days						
6th	0.3707	0.0870	–	–	–	–
7th	0.3971	0.2160	0.1290	0.1627	0.2344	0.1054
8th	0.6490	0.2970	0.0810	0.2447	0.4043	0.3233
9th	0.4174	0.4560	0.1590	0.4457	–0.0283	–0.1873
Second moult period of 2 days						
10th	0.5105	0.4513	–	–	–	–
11th	0.9247	0.5723	0.1210	0.5917	0.3330	0.2120
12th	1.5542	0.8757	0.3033	0.8647	0.6895	0.3862
13th	1.6204	1.3223	0.4466	1.0190	0.6014	0.1546
Third moult period of 3 days						
14th	2.7208	2.0447	–	–	–	–
15th	1.9023	2.8173	0.7727	0.9233	0.9790	0.2064
16th	4.1137	3.1797	0.3623	1.0573	3.0564	2.6940
17th	3.1915	3.5647	0.3850	1.0303	2.1612	1.7772
18th	4.0555	4.2450	0.6803	1.1767	2.8788	2.1985
19th	3.6477	4.9760	0.7310	1.1747	2.4730	1.7420
Fourth moult period of 3 days						
20th	2.9834	6.7510	–	–	–	–
21st	2.9980	7.3610	0.6100	0.8183	2.1797	1.5697
22nd	6.3581	8.4740	1.1130	1.6150	4.7431	3.6301
23rd	6.5401	8.8746	0.4006	2.5650	3.9751	3.5744
24th	5.7489	8.8513	–0.0233	2.3631	3.3858	3.4091
25th	6.9657	9.5175	0.6662	2.7776	4.1881	3.5219
26th	8.7554	13.6351	4.1176	2.8018	5.9537	1.8361

g, which was the last day of the third instar. Irregular trend was observed from fifteenth day 0.9233 g to ninetieth day 1.1747 g. In fifth instar, the weight of litter increased till twenty-sixth day 2.8018 g, and twenty third day 2.5650 g which was the maximum weight of litter excreted and then decreased till twenty fourth day 2.3631 g. On the last larval day, *i.e.* twenty sixth days, the weight of litter excreted was 2.8018 g.

Food Energy Assimilation (FEA)

FEA was 0.0694 g on second day. It decreased to 0.0524 g on fourth day of first instar. In second instar, it was 0.2344 g and 0.4043 g on seventh day and eighth day. However, it decreased to –0.0283 g on the last day of the second instar showing negative value. In third instar, the FEA was maximum 0.6895 g on twelfth day. FEA did not showed any regular trend in fourth and fifth instar. It was 3.0564 g on sixteenth day and 4.7431 g on twenty second day. Maximum value of FEA was observed on twenty-sixth day 5.9537 g.

Respiration (R)

Respiration value was calculated to be in decrease on the last day of first instar –0.0008 g, second instar –0.1873 g and on last two days of third instar 0.3862 g and 0.1546g. Fourth instar of second day 0.2064g, fifth instar positive respiration was recorded on rest of the feeding days.

Daily Food Consumption and Utilization per Larva of *A. selene* reared on *T. arjuna*

Food Consumption (F)

A larva of *A.selene* fed on Arjuna consumed 0.0381g of leaf on first day, 0.0929 g on second day, 0.0650 g on third day, 0.0510 g on fifth day of feeding. Decline in food consumption was observed on the fifth day, which was the last day of the first instar. There was progressive increase in food consumption from sixth day 0.3770g to eighth day 0.6375 g, from tenth day 0.9226 g to fifteenth day 4.9966 g and sixteenth day 3.2366 g to eighteenth day 4.1408 g of feeding. From twenty first day 11.5258 g to thirty sixth day 7.9364 g food consumption was irregular. Maximum food was consumed on thirty fifth day 13.4785 g.

Table 9: Daily Food Consumption and Utilization Per Larva of *A.selene* reared on *T. arjuna*

Unit: g.

Day of Feeding	*F*	*A*	*G*	*E*	*FEA*	*R*
1st	0.0381	0.0083	–	–	–	–
2nd	0.0929	0.0133	0.0051	0.0143	0.0786	0.0736
3rd	0.0650	0.0224	0.0091	0.0636	0.0014	–0.0076
4th	0.0545	0.0601	0.0377	0.0725	–0.0180	–0.0558
5th	0.0510	0.0773	0.0172	0.0136	0.0374	0.0202
		First moult period of 1 days				
6th	0.3770	0.0787	–	–	–	–
7th	0.4035	0.2260	0.1473	0.1433	0.2602	0.1129
8th	0.6375	0.3573	0.1313	0.1830	0.4545	0.3232
9th	0.4178	0.4743	0.1170	0.4100	0.0078	–0.1092
10th	0.9226	0.8837	0.4094	0.6190	0.3036	–0.1058
11th	1.1697	1.1480	0.2643	0.5923	0.5874	0.3231
		Second moult period 2 days				
12th	2.1372	2.1327	–	–	–	–
13th	3.1568	2.6550	0.5223	1.1230	2.0338	1.5115
14th	4.5755	3.9913	1.3363	1.3150	3.2605	1.9241
15th	4.9966	5.2020	1.2107	1.4053	3.5913	2.3806
16th	3.2366	6.2673	1.0653	1.4400	1.7966	0.7313
17th	4.6662	6.7900	0.5227	0.8353	3.8308	3.3082
18th	4.1408	9.5040	2.7140	2.4133	1.7275	–0.9865
		Third moult period 3 days				
19th	8.5331	12.2943	–	–	–	–
20th	10.1219	14.7030	2.4087	5.1523	4.9696	2.5609
21st	11.5258	17.8557	3.1527	7.0150	4.5108	1.3582
22nd	9.4984	20.2467	2.3910	7.0427	2.4557	0.0647
23rd	11.4984	22.6453	2.3987	8.5407	2.9577	0.5590
24th	9.3565	23.0650	0.4197	7.3090	2.0475	1.6278
25th	8.6934	22.9973	–0.0677	7.1200	1.5734	1.6411

Contd...

Table 9–Contd...

Day of Feeding	*F*	*A*	*G*	*E*	*FEA*	*R*
		Fourth moult period 3 days				
26th	9.4984	20.2467	–	–	–	–
27th	11.4984	22.6453	2.3987	8.5407	2.9577	0.5590
28th	9.3565	23.0650	0.4197	7.3090	2.0475	1.6278
29th	8.6934	22.9973	–0.0677	7.1200	1.5734	1.6411
30th	8.0942	23.0563	0.0591	6.1407	1.9535	1.8945
31st	7.3279	22.8927	–0.1637	4.9308	2.3971	2.5608
32nd	10.0351	22.1838	–0.7088	5.7864	4.2487	4.9575
33rd	9.2133	23.8950	1.7112	7.2349	1.9784	0.2673
34th	13.9212	24.6611	0.5539	8.3328	5.5884	5.0345
35th	13.4785	25.0083	0.3472	8.7217	4.7568	4.4096
36th	7.9364	23.3663	–1.9771	4.5946	3.3418	5.3189

Mean Weight of Larva (A)

Weight of freshly hatched larva of *A.selene* was 0.0083g. On subsequent days it increased progressively till thirty sixth days when the larva attained the maximum weight of 23.8950 g in fifth instar. Larval weight on day fifth, ninth, tenth, fifteenth, sixteenth, twenty-first, twenty second and thirty-sixth day was 0.0773 g, 0.4743g, 0.8837 g, 5.2020 g, 6.2673 g, 17.8557 g, 20.2467 g and 23.3663 g respectively.

Weight Gain (G)

Initial weight gain in larva was calculated to be 0.0051g on second day. It increased daily till fourth day 0.0377 g and decreased on fifth day 0.0172 g which was the last day of the first instar. Irregular growth pattern was observed, second instar to last instar, on twenty fifth days –0.0677 g, Negative weight gain was observed on twenty-fifth day. Maximum weight gain was calculated on twenty first day 3.1527g which was third day of fourth instar and minimum on twenty-fifth day –0.0677 g indicating negative scope for growth during this period.

Excreta (E)

Weight of litter on second day was found to be 0.0143 g. It progressively increased on subsequent days till tenth day 0.6190 g, which was the fifth day of the second instar. Irregular trend was observed from fifteenth day 1.4053 g to thirty sixth day 4.5946 g. In fifth instar, the weight of litter increased till thirty fifth day 8.7217 g, and thirty fifth days 8.7217 which was the maximum weight of litter excreted and then decreased till thirty sixth day 4.5946 g. which is last larval day, *i.e.* thirty sixth days.

Food Energy Assimilation (FEA)

FEA was 0.0786 g on second day. It decreased to 0.0014 g on third day of first instar. In second instar, it was 0.2602 g and 0.4545 g on seventh day and eighth day. In third instar, the FEA was maximum 3.8308 g on seventieth day. FEA did not show any regular trend in fourth and fifth instar. It was 4.9696 g on twenty day and 2.0475 g on twenty eighth days. Maximum value of FEA was observed on thirty fourth day 5.5884 g.

Respiration (R)

Respiration value was calculated to be in decrease on the last day of first instar 0.0202 g, second instar 0.3231 g and on last two days of third instar 3.3082 g and –0.9865g. Fourth instar of second day 2.5609 g, fifth instar positive respiration was recorded on rest of the feeding days.

Daily Food Consumption and Utilization per Larva of *A. selene* reared on *T. tomentosa*

Food Consumption (F)

A larva of *A.selène* fed on Ain consumed 0.0448g of leaf on first day, 0.0967 g on second day, 0.0922 g on third day, 0.2402 g on fifth day of feeding. Decline in food consumption was not observed in the first instar. There was progressive increase in food consumption from sixth day 0.3648g to tenth day 0.9092 g, from twelfth day 0.7913 g to fifteenth day 1.8845 g and seventieth day 2.2028 g to eighteenth day 2.4619 g of feeding. From twenty first day 4.3852 g to thirty sixth day 6.4039 g food consumption was irregular. Maximum food was consumed on thirty fifth day 7.9721 g.

Mean Weight of Larva (A)

Weight of freshly hatched larva of *A.selene* was 0.0107g. On subsequent days it increased progressively till thirty sixth days when the larva attained the maximum weight of 18.1507 g in fifth instar.

Table 10: Daily Food Consumption and Utilization Per Larva of *A.selene* reared on *T. tomentosa* (Ain)

Unit: g.

Day of Feeding	F	A	G	E	FEA	R
1st	0.0448	0.0107	–	–	–	–
2nd	0.0967	0.0145	0.0039	0.0105	0.0862	0.0823
3rd	0.0922	0.0363	0.0217	0.0191	0.0731	0.0514
4th	0.1359	0.0644	0.0281	0.0341	0.1018	0.0736
5th	0.2409	0.0867	0.0223	0.2002	0.0407	0.0184
		First moult period of 1 days				
6th	0.3648	0.1713	–	–	–	–
7th	0.3860	0.3077	0.1363	0.1563	0.2297	0.0933
8th	0.4805	0.3840	0.0763	0.2233	0.2572	0.1809
9th	0.7275	0.7477	0.1523	0.2573	0.4701	0.3178
10th	0.9092	0.8173	0.0697	0.1720	0.7372	0.6675
11th	0.1779	1.0067	0.1893	0.1683	0.0095	–0.1798
		Second moult period of 2 days				
12th	0.7913	1.6567	–	–	–	–
13th	1.7183	1.9873	0.3307	0.2870	1.4313	1.1006
14th	1.8440	2.3423	0.3550	0.4430	1.4010	1.0460
15th	1.8845	2.7557	0.4133	0.4980	1.3865	0.9732
16th	1.5347	3.2540	0.4983	0.6067	0.9281	0.4297
17th	2.2028	3,8577	0.6037	0.6143	1.5885	0.9848
18th	2.4619	4.6497	0.7920	0.7070	1.7549	0.9629
		Third moult period of 3 days				
19th	3.3810	10.0867	–	–	–	–
20th	3.5585	10.9330	0.8463	1.6700	1.8885	1.0422
21st	4.3852	11.8530	0.9200	2.3040	2.0812	1.1612
22nd	3.9767	12.7960	0.9430	1.9593	2.0174	1.0744
23rd	4.7088	13.9333	1.1373	2.4330	2.2758	1.1385
24th	5.5583	13.1350	–0.7983	3.3257	2.2326	1.4343
25th	6.3480	14.9880	1.8530	4.2560	2.0920	0.2390

Contd...

Table 10–Contd...

Day of Feeding	*F*	*A*	*G*	*E*	*FEA*	*R*
		Fourth moult period of 3 days				
26th	5.8376	14.7993	–	–	–	–
27th	6.7333	15.4703	0.6710	3.8237	2.9096	2.2386
28th	6.6010	15.0430	–0.4273	3.5203	3.0806	2.6534
29th	5.8666	16.0593	1.0163	3.2887	2.5779	1.5616
30th	4.6069	16.9101	0.8508	2.4670	2.1399	1.2891
31st	4.5586	16.9116	0.0015	1.9986	2.5600	2.5585
32nd	3.5389	16.9098	–0.0018	2.6782	0.8607	0.8589
33rd	3.6720	17.4837	0.5739	2.3731	1.2989	0.7251
34th	6.5198	17.5683	0.0846	2.5454	3.9745	3.8898
35th	7.9721	17.6618	0.0935	2.1497	5.8224	5.7288
36th	6.4039	18.1507	0.4889	1.8519	4.5519	4.0631

Larval weight on day fifth, ninth, tenth, fifteenth, sixteenth, twenty-first, twenty second and thirty-sixth day was 0.0867 g, 0.7477g, 0.8173 g, 2.7557 g, 3.2540 g, 11.8530 g, 12.7960 g and 18.1507 g respectively.

Weight Gain (G)

Initial weight gain in larva was calculated to be 0.0039g on second day. It increased daily till fourth day 0.0281 g and decreased on fifth day 0.0223 g which was the last day of the first instar. Irregular growth pattern was observed, second instar to last instar, on twenty fourth days –0.7983 g, Negative weight gain was observed on twenty-fourth day. Maximum weight gain was calculated on twenty fifth day 1.8530 g which was last day of fourth instar and minimum on thirty second day –0.0018 g indicating negative scope for growth during this period.

Excreta (E)

Weight of litter on second day was found to be 0.0105 g. It progressively increased on subsequent days till ninth day 0.2573 g, which was the fourth day of the second instar. Irregular trend was observed from fifteenth day 0.4980 g to thirty sixth days 1.8519 g. In fifth instar, Irregular trend was observed,the weight of litter increased till thirty fifth day 2.1497 g, and twenty seventh day 3.8237g which was the maximum weight of litter excreted and then decreased till thirty sixth day 1.8519 g. which is last larval day, *i.e.* thirty sixth days.

Food Energy Assimilation (FEA)

FEA was 0.0862 g on second day. It decreased to 0.0731g on third day of first instar. In second instar, it was 0.2297 g and 0.2572 g on seventh day and eighth day. In third instar, the FEA was maximum 1.7549 g on eighteenth day. FEA did not show any regular trend in fourth and fifth instar. It was 2.0174 g on twenty day and 3.0806 g on twenty eighth days. Maximum value of FEA was observed on thirty fifth day 5.8224 g.

Respiration (R)

Respiration value was calculated to be in decrease on the last day of first instar 0.0184 g, second instar –0.1798 g and on last two days of third instar 0.9848 g and 0.9629g. Fourth instar of second day 1.0422 g, fifth instar positive respiration was recorded on rest of the feeding days.

Daily Food Consumption and Utilization per Larva of *A. selene* reared on *T. catappa*

Food Consumption (F)

A larva of *A.selene* fed on Badam consumed 0.0081g of leaf on first day, 0.0345 g on second day, 0.0428 g on third day, 0.0890 g on fifth day of feeding. Increasing in food consumption was observed on the first day to ninenth day of the second instar. There was progressive increase in food consumption from sixth day 0.0421 g to eighth day 0.2167 g, from ninenth day 0.0.2423 g, after fifteenth day to thirty sixth day irregular of feeding. From thirty second day 2.2200 g to thirty sixth day 6.9000 g food consumption was irregular. Maximum food was consumed on thirty sixth day 6.9000 g.

Mean Weight of Larva (A)

Weight of freshly hatched larva of *A.selene* was 0.0067g. On subsequent days it increased progressively till thirty sixth days when the larva attained the maximum weight of 10.4800 g in fifth instar. Larval weight on day fifth, ninth, tenth, fifteenth, sixteenth, twenty-

Table 11: Daily Food Consumption and Utilization Per Larva of *A.selene* reared on *T. catappa* (Badam)

Unit: g.

Day of Feeding	*F*	*A*	*G*	*E*	*FEA*	*R*
1st	0.0081	0.0067	–	–	–	–
2nd	0.0345	0.0111	0.0044	0.0205	0.0140	0.0096
3rd	0.0428	0.0257	0.0146	0.0320	0.0108	–0.0038
4th	0.0578	0.0476	0.0219	0.0369	0.0209	–0.0010
5th	0.0890	0.0558	0.0082	0.0071	0.0818	0.0736
First moult period of 1 days						
6th	0.0421	0.0616	–	–	–	–
7th	0.1465	0.0760	0.0144	0.0604	0.0861	0.0717
8th	0.2167	0.1104	0.0344	0.1036	0.1131	0.0787
9th	0.2423	0.1960	0.0856	0.1240	0.1183	0.0327
10th	0.1266	0.2600	0.0640	0.0496	0.0770	0.0130
11th	0.1743	0.2400	–0.0200	0.0444	0.1299	0.1499
Second moult period 2 days						
12th	0.2921	0.2844	–	–	–	–
13th	0.2864	0.2896	0.0052	0.0532	0.2332	0.2280
14th	0.3596	0.3528	0.0632	0.1040	0.2556	0.1924
15th	0.3058	0.4224	0.0696	0.1400	0.1658	0.0962
16th	0.2903	0.4964	0.0740	0.1320	0.1583	0.0843
17th	0.4731	0.6152	0.1188	0.1800	0.2931	0.1743
18th	0.3946	0.8192	0.2040	0.2572	0.1374	–0.0667
Third moult period of 3 days						
19th	0.5388	1.1872	–	–	–	–
20th	0.4228	1.2736	0.0864	0.2000	0.2228	0.1364
21st	0.4504	1.6048	0.3312	0.2728	0.1776	–0.1536
22nd	1.2223	1.7928	0.1880	0.3904	0.8319	0.6439
23rd	1.2216	2.0160	0.2232	0.4696	0.7520	0.5288
24th	1.8455	2.3520	0.3360	0.5040	1.3415	1.0055
25th	0.8389	2.5440	0.1920	0.4202	0.4188	0.2268

Contd...

Table 11–Contd...

Day of Feeding	*F*	*A*	*G*	*E*	*FEA*	*R*
			Fourth moult period of 3 days			
26th	3.2600	4.0500	–	–	–	–
27th	3.7200	5.6000	1.5500	0.2080	3.5120	1.9620
28th	2.1000	6.7300	1.1300	0.2500	1.8500	0.7200
29th	2.5500	6.3200	–0.4100	0.3520	2.1980	1.7850
30th	3.0500	4.1800	–2.1400	1.1015	1.9485	–0.1915
31st	2.4800	7.5000	3.3200	1.6215	0.8285	–2.4915
32nd	2.2000	6.1000	–1.4000	1.4220	0.7780	–0.6220
33rd	3.9500	6.7300	0.6300	1.4218	2.2582	1.8982
34th	4.5600	8.5000.	1.7700	1.6522	2.9078	1.1378
35th	5.8200	9.1000	0.6000	0.8990	4.9210	4.3210
36th	6.9000	10.4800	1.3800	1.6844	5.2156	3.8356

first, twenty second and thirty-sixth day was 0.0558 g, 0.1960g, 0.2600 g, 0.4224 g, 0.4964 g, 1.6048 g, 1.7978 g and 10.4800 g respectively.

Weight Gain (G)

Initial weight gain in larva was calculated to be 0.0044g on second day. It increased daily till fourth day 0.0219 g and decreased on fifth day 0.0082g which was the last day of the first instar. Irregular growth pattern was observed, second instar to last instar, on twenty fifth days 0.1920 g, Negative weight gain was observed on eleventh day. Maximum weight gain was calculated on thirty first day 3.3200 g which was sixth day of fifth instar and minimum on twenty-ninenth day –0.4100 g indicating negative scope for growth during this period.

Excreta (E)

Weight of litter on second day was found to be 0.0205 g. It progressively increased on subsequent days till ninenth day 0.1240 g, which was the fourth day of the second instar. Irregular trend was observed from fifteenth day 0.1400 g to thirty sixth days 1.6844 g. In fifth instar, the weight of litter increased till twenty seventh day 0.2080 g, to thirty first days 1.6215g, thirty sixth day 1.6844g which was the maximum weight of litter excreted.

Food Energy Assimilation (FEA)

FEA was 0.0140 g on second day. It decreased to 0.0770 g on tenth day it is sixth day of second instar. In second instar, it was 0.0861 g and 0.1131 g on seventh day and eighth day. In third instar, the FEA was maximum 0.2931 g on seventieth day. FEA did not show any regular trend in fourth and fifth instar. It was 0.8319 g on twenty day and 1.8500 g on twenty eighth days. Maximum value of FEA was observed on thirty sixth day 5.2156 g.

Respiration (R)

Respiration value was calculated to be in decrease on the last day of first instar 0.0736 g, second instar 0.1499 g and on last two days of third instar 0.1743 g and –0.0667g. Fourth instar of second day 0.1364 g, fifth instar positive respiration was recorded on rest of the feeding days.

Daily Food Consumption and Utilization per Larva of *A. mylitta kolhapurensis* Reared on *T. catappa*

Food Consumption (F)

A larva of *A. mylitta kolhapurensis* fed on Badam consumed 0.0542 g of leaf on first day, 0.0762 g on second day, 0.0868 g on third day, 0.0841 g on fourth day of feeding. Decline in food consumption was observed on the fourth day, which was the last day of the first instar. There was progressive increase in food consumption from fifth day 0.3207 g to seventh day 0.6498 g from ninth day 0.5362 g to thirteen day 1.9309 g and fifteenth day 2.7208 g to nineteen day 5.6818 g of feeding. From twenty first day 6.4211 g to thirteenth day 12.4249 g food consumption was irregular. Maximum food was consumed on twenty-eight day 14.1210 g.

Mean Weight of Larva (A)

Weight of freshly hatched larva of *A. mylitta kolhapurensis* was 0.0082g. On subsequent days it increased progressively till twenty-seventh day when the larva attained the maximum weight of 19.2111 g in fifth instar. Larval weight on day fifth, ninth, tenth, fifteenth, sixteenth, twenty-first, twenty second and twentyninenth day was 0.0871 g, 0.4610 g, 0.5712 g, 2.0105 g, 2.6812 g, 7.0103g, 8.1242 g and 18.8414 g respectively.

Table 12: Daily Food Consumption and Utilization per Larva of *A. mylitta kolhapurensis* Reared on *Terminalia catappa* during– First Crop (June–July)

Unit: g.

Day of Feeding	*F*	*A*	*G*	*E*	*FEA*	*R*
1st	0.0542	0.0082	–	–	–	–
2nd	0.0762	0.0131	0.0049	0.0106	0.0656	0.0607
3rd	0.0868	0.0272	0.0141	0.0193	0.0675	0.0534
4th	0.0841	0.0619	0.0347	0.0340	0.0501	0.0154
		First moult period of 1.5 days				
5th	0.3207	0.0871	–	–	–	–
6th	0.3472	0.2165	0.1294	0.0892	0.2580	0.1286
7th	0.6498	0.2970	0.0805	0.1065	0.5433	0.4628
8th	0.5190	0.4660	0.1690	0.1244	0.3946	0.2256
		Second moult period of 2 days				
9th	0.5362	0.4610	–	–	–	–
10th	0.9861	0.5712	0.1102	0.5642	0.4219	0.3117
11th	1.6542	0.8977	0.3265	0.5210	1.1332	0.8067
12th	1.7208	1.4232	0.5255	1.0127	0.7081	0.1826
13th	1.9309	1.9772	0.5540	0.8928	1.0381	0.4841
14th	1.6503	1.9968	0.0196	1.0017	0.6486	0.6290
		Third moult period of 3days				
15th	2.7208	2.0105	–	–	–	–
16th	4.1138	2.6812	0.6707	0.9264	3.1874	2.5167
17th	4.4241	3.0407	0.3595	1.9211	2.5030	2.1435
18th	4.4549	4.0070	0.9663	1.8942	2.5607	1.5944
19th	5.6818	4.6760	1.6690	2.1021	3.5797	1.9107
20th	4.6710	6.7169	1.0409	2.0107	2.6603	1.6194
		Fourth moult period of 4 day				
21st	6.4211	7.0103	–	–	–	–
22nd	7.6998	8.1242	1.1139	1.4211	6.2787	5.1648
23rd	8.1211	9.1649	1.0407	2.6879	6.4332	4.3925
24th	10.6842	11.2023	2.0374	4.7253	5.9589	3.9215

Contd...

Table 12–Contd...

Day of Feeding	*F*	*A*	*G*	*E*	*FEA*	*R*
25th	12.4216	14.2241	5.0574	4.9248	7.4968	2.4394
26th	11.6889	16.3909	2.1668	6.1911	5.4978	3.3310
27th	12.4211	19.2111	2.8202	8.1242	4.2969	1.4767
28th	14.1210	17.9892	–1.3119	7.1410	6.9800	8.2919
29th	13.1118	18.8414	0.9422	7.0109	6.1009	5.1587
30th	12.4249	17.4211	–1.4203	8.1021	4.3228	6.5958

Weight Gain (G)

Initial weight gain in larva was calculated to be 0.0049 g on second day. It increased daily till sixth day 0.1294 g and decreased on seventh day 0.0805 g which was the third day of the second instar. In third instar also the weight gain decreased on the last day only *i.e.* on fourteenth day 0.0196 g Weight gain again increased from twenty-second day 1.0407 g up to twenty-fifth day 5.0574 g Afterwards decrease in weight gain was observed till twenty-sixth day. Negative weight gain was observed on twenty-eighth day. Maximum weight gain was calculated on twenty-fifth day 5.0574g and minimum on twenty-eight day –1.3119 g indicating negative scope for growth during this period.

Excreta (E)

Weight of litter on second day was found to be 0.0106 g. It progressively increased on subsequent days till tenth day 0.5642 g, which was the second day of the third instar. Daily increase was observed from sixteen day 0.9264 g to nineteen day 2.1021 g followed by a decrease on twenty day 2.0107 g which was fourth instar. In fifth instar, the weight of litter increased till twenty-seventh day 8.1242 g, which was the maximum weight of litter excreted and then decreased till twenty ninenth day 7.0109 g. On the last larval day of fifth instar, 8.1021g.

Food Energy Assimilation (FEA)

FEA was 0.0656 g on second day. It decreased to 0.0501 g on fourth day of first instar. In second instar, it was 0.2580 g and 0.5433 g on sixth day and seventh day. However, it decreased to 0.3946 g on the last day of the second instar. In third instar, the FEA was maximum 1.1332 g on eleventh day. In fourth instar, it increased till nineteen day 3.5797 g, decreased on the following day 2.6603 g. FEA did not show any regular trend in fifth instar. It was 6.2787 g on twenty-second day and 4.3228 g on thirteenth day. Maximum value of FEA was observed on twenty-fifth day 7.4978 g.

Respiration (R)

Respiration value was calculated to be in decrease on the last day of first instar 0.0154 g, second instar 0.2256 g and on last two days of third instar –0.1826 g. and positive highest value was also recorded on twenty-eight day 8.2919 g, twenty-fourth day 3.9215 g and twenty-sixth day 3.3310 g during fifth instar. Positive respiration was recorded on rest of the feeding days.

Daily Food Consumption and Utilization per Larva of *A. mylitta kolhapurensis* Reared on *T. catappa*

Food Consumption (F)

A larva of *A. mylitta kolhapurensis* fed on Badam consumed 0.0462 g of leaf on first day, 0.0671 g on second day, 0.0842 g on third day, 0.0760 g on fourth day of feeding. Decline in food consumption was observed on the fourth day, which was the last day of the first instar. There was progressive increase in food consumption from fifth day 0.2102 g to seventh day 0.6842 g from ninth day 0.5269 g to twelfth day 2.1242 g and fifteenth day 2.4211 g to nineteen day 6.8994 g of feeding. From twenty first day 5.8992 g to thirteenth day 13.6542 g food consumption was irregular. Maximum food was consumed on twenty-seventh day 14.4211 g.

Mean Weight of Larva (A)

Weight of freshly hatched larva of *A. mylitta kolhapurensis* was 0.0083g. On subsequent days it increased progressively till twenty-seventh day when the larva attained the maximum weight of 19.0262 g in fifth instar. Larval weight on day fifth, ninth, tenth, fifteenth, sixteenth, twenty-first, twenty second and twentyninenth day was 0.0492g, 0.4821 g, 1.8942 g, 2.1211g, 7.8991g, 8.1214 g and 18.0421 g respectively.

Weight Gain (G)

Initial weight gain in larva was calculated to be 0.0025 g on

Table 13: Daily Food Consumption and Utilization per Larva of *A. mylitta kolhapurensis* Reared on *Terminalia catappa* during– Second Crop (Sep–Oct)

Unit: g.

Day of Feeding	*F*	*A*	*G*	*E*	*FEA*	*R*
1st	0.0462	0.0083	–	–	–	–
2nd	0.0671	0.0108	0.0025	0.0112	0.0559	0.0534
3rd	0.0842	0.0242	0.0134	0.0205	0.0637	0.0503
4th	0.0760	0.0465	0.0223	0.0415	0.0345	0.0122
	First moult period of 1.5 days					
5th	0.2101	0.0492	–	–	–	–
6th	0.3213	0.0949	0.0457	0.0452	0.2761	0.02304
7th	0.6842	0.2231	0.1282	0.0892	0.5950	0.4668
8th	0.5142	0.4749	0.2518	0.1820	0.3322	0.0804
	Second moult period of 2 days					
9th	0.5269	0.4821	–	–	–	–
10th	0.8942	0.6992	0.2171	0.2011	0.6931	0.4760
11th	1.1021	0.8942	0.1950	0.5219	0.5802	0.3852
12th	2.1242	1.1211	0.2269	1.1122	1.0120	0.7851
13th	1.9308	1.4268	0.3057	0.9924	0.9384	0.6327
14th	1.8704	1.6978	0.2710	1.0201	0.8503	0.5793
	Third moult period of 3days					
15th	2.4211	1.8942	–	–	–	–
16th	4.6068	2.1211	0.2269	0.9764	3.6304	3.4035
17th	4.6542	4.6813	2.5602	1.3712	3.2830	0.7228
18th	5.6121	5.1289	0.4476	1.9891	3.6230	3.1754
19th	6.8994	6.4169	1.2880	2.3211	3.3600	2.1566
20th	5.6811	7.6203	1.2034	2.3211	3.3600	2.1566
	Fourth moult period of 4 day					
21st	5.8992	7.8991	–	–	–	–
22nd	6.1213	8.1214	0.2223	1.6839	4.4374	4.2151
23rd	8.6811	10.0018	1.8804	3.7211	4.9600	3.0796
24th	8.4214	11.0107	1.0089	4.2210	5.2004	3.1915

Contd...

Table 13–Contd...

Day of Feeding	*F*	*A*	*G*	*E*	*FEA*	*R*
25th	10.4210	13.0542	2.0435	4.8924	5.5286	3.4851
26th	12.6089	17.1169	4.0627	6.7874	5.8215	1.7588
27th	14.4211	19.0262	1.9093	8.4119	6.0092	4.0999
28th	13.4019	17.0680	–1.9582	6.6282	6.7737	8.7319
29th	13.8989	18.0421	0.9741	7.1021	6.7968	5.8227
30th	13.6542	18.0122	–0.0299	8.7927	4.8615	4.8914

second day. It increased daily till tenth day 0.2171 g and decreased on eleventh day 0.1950 g which was the third day of the third instar. In third instar also the weight gain decreased on the last day only *i.e.* on fourteenth day 0.2710 g, Weight gain again decreased on the eighteen day 0.4476 g, twenty-second day 0.2223 g, Afterwards increase in weight gain was observed till twenty-sixth day 4.0627. Negative weight gain was observed on twenty-eighth day. Maximum weight gain was calculated on twenty-sixth day 4.0627g and minimum on twenty-eight day –1.9582 g indicating negative scope for growth during this period.

Excreta (E)

Weight of litter on second day was found to be 0.0112 g. It progressively increased on subsequent days till twelfth day 1.1122 g, which was the third day of the third instar. Daily increase was observed from seventeen day 1.3712 g to nineteen day 2.6412 g followed by a decrease on twenty day 2.3211 g which was fourth instar. In fifth instar, the weight of litter increased till twenty-seventh day 8.4219 g, which was the maximum weight of litter excreted and then decreased till twenty ninenth day 6.6282 g. On the last larval day of fifth instar, 8.7927g.

Food Energy Assimilation (FEA)

FEA was 0.0559 g on second day. It decreased to 0.0345 g on fourth day of first instar. In second instar, it was 0.2761 g and 0.5950 g on sixth day and seventh day. However, it decreased to 0.3322 g on the last day of the second instar. In third instar, the FEA was maximum 1.1122 g on twelfth day. In fourth instar, it increased till nineteen day 4.2582 g, decreased on the following day 3.3600 g. FEA did not show any regular trend in fifth instar. It was 4.4374 g on twenty-second day and 4.8615 g on thirteenth day. Maximum value of FEA was observed on twenty-ninenth day 6.7968 g.

Respiration (R)

Respiration value was calculated to be in decrease on the last day of first instar 0.0122 g, second instar 0.0804 g and on last two days of third instar 0.5793 g. and positive highest value was also recorded on twenty-eight day 8.7319g, twenty-fourth day 3.1915g and twenty-sixth day 1.7588 g during fifth instar. Positive respiration was recorded on rest of the feeding days.

Daily Food Consumption and Utilization per Larva of *A. mylitta kolhapurensis* Reared on *T. arjuna*

Food Consumption (F)

A larva of *A. mylitta kolhapurensis* fed on *arjuna* consumed 0.0392 g of leaf on first day, 0.0850 g on second day, 0.0740 g on third day, 0.0445 g on fourth day of feeding. Decline in food consumption was observed on the fourth day, which was the last day of the first instar. There was progressive increase in food consumption from fifth day 0.3645 g to seventh day 0.6057 g, from ninth day 0.5165 g to thirteen day 1.4909 g and fifteenth day 2.1392 g to eighteen day 4.9988 g of feeding. From twenty first day 4.4218 g to thirteenth day 11.7985 g food consumption was irregular. Maximum food was consumed on twenty-ninenth day 14.9214 g.

Mean Weight of Larva (A)

Weight of freshly hatched larva of *A. mylitta kolhapurensis* was 0.0083g. On subsequent days it increased progressively till twenty-seventh day when the larva attained the maximum weight of 23.0850 g in fifth instar. Larval weight on day fifth, ninth, tenth, fifteenth, sixteenth, twenty-first, twenty second and twentyninenth day was 0.0642 g, 0.4685 g, 0.8890 g, 2.2618 g, 2.6551 g, 9.4050g, 10.3156 g and 23.3042 g respectively.

Weight Gain (G)

Initial weight gain in larva was calculated to be 0.0037g on second day. It increased daily till sixth day 0.0.919 g and decreased

Plate 18: Rearing of *Anthemea mylitta* on *T. catappa*

Figure 75: Rearing of 1st Instars in Plastic Cage; Figure 76: Rearing of 2nd Instars in Plastic Tray; Figure 77: Rearing of 3rd Instars in G.I. Tray; Figure 78: Rearing of 4th Instars in G.I. Tray; Figure 79: Rearing of 5th Instars in G.I. Tray; Figure 80: Cocoons

Plate 19: Rearing of *Anthemea mylitta* on *T. arjuna*

Figure 81: Rearing of 1st Instars in Plastic Cage; Figure 82: Rearing of 2nd Instars in Plastic Tray; Figure 83: Rearing of 3rd Instars in G.I. Tray; Figure 84: Rearing of 4th Instars in G.I. Tray; Figure 85: Rearing of 5th Instars in G.I. Tray; Figure 86: Cocoons

Plate 20: Rearing of *Antheraea mylitta* on *T. tomentosa*

Figure 87: Rearing of 1st Instars in Plastic Cage; Figure 88: Rearing of 2nd Instars in Plastic Tray; Figure 89: Rearing of 3rd Instars in G.I. Tray; Figure 90: Rearing of 4th Instars in G.I. Tray; Figure 91: Rearing of 5th Instars in G.I. Tray; Figure 92: Cocoons

Plate 21: Rearing of *Antheraea mylitta* on *Z. jujuba*

Figure 93: Rearing of 1st Instars in Plastic Tray; Figure 94: Rearing of 2nd Instars in Plastic Tray; Figure 95: Rearing of 3rd Instars in G.I. Tray; Figure 96: Rearing of 4th Instars in G.I. Tray; Figure 97: Rearing of 5th Instars in G.I. Tray; Figure 98: Cocoons.

on seventh day 0.0699 g which was the third day of the second instar. In third instar also the weight gain decreased on thirteenth day 0.0115 g, Weight gain again increased from twenty-second day 0.9106 g up to twenty-third day 4.3869 g Afterwards decrease in weight gain was observed till twenty-sixth day. Negative weight gain was observed on twenty-eighth day. Maximum weight gain was calculated on twenty-third day 4.3869 g and minimum on twenty-eight day –0.1922 g indicating negative scope for growth during this period.

Excreta (E)

Weight of litter on second day was found to be 0.0216 g. It progressively increased on subsequent days till twelfth day 1.0212 g, which was the fourth day of the third instar. Daily increase was observed from sixteen day 0.9852 g to nineteen day 2.4218 g followed by a decrease on twenty day 2.4012 g which was fourth instar. In fifth instar, the weight of litter increased till twenty-seventh day 8.1109 g, which was the maximum weight of litter excreted and then decreased till twenty eighth day 7.0014 g,the last larval day of fifth instar, 8.9029g.

Food Energy Assimilation (FEA)

FEA was 0.0634 g on second day. It decreased to 0.0093 g on fourth day of first instar. In second instar, it was 0.2990 g and 0.4889 g on sixth day and seventh day. However, it decreased to 0.3193 g on the last day of the second instar. In third instar, the FEA was maximum 0.6933 g on eleventh day. In fourth instar, it increased till eighteenth day 3.1072g, decreased on the following day 1.4886 g. FEA did not show any regular trend in fifth instar. It was 7.9088 g on twenty ninenth day and 2.8956 g on thirteenth day. Maximum value of FEA was observed on twenty ninenth day 7.9088 g.

Respiration (R)

Respiration value was calculated to be in decrease on the last day of first instar –0.0211 g, second instar 0.1889 g and on last two days of third instar –0.7214 g. and positive highest value was also recorded on twenty ninenth day 7.4974 g, twenty-fourth day 3.4708 g and twenty-sixth day 1.2769 g during fifth instar. Positive respiration was recorded on rest of the feeding days.

Table 14: Daily Food Consumption and Utilization per Larva of *A. mylitta kolhapurensis* reared on *Terminalia arjuna* during– First Crop (June–July)

Unit: g.

Day of Feeding	*F*	*A*	*G*	*E*	*FEA*	*R*
1st	0.0392	0.0083	–	–	–	–
2nd	0.0850	0.0120	0.0037	0.0216	0.0634	0.0597
3rd	0.0740	0.0214	0.0094	0.0484	0.0256	0.0162
4th	0.0445	0.518	0.0304	0.0352	0.0093	–0.0211
	First moult period of 1.5 days					
5th	0.3645	0.0642	–	–	–	–
6th	0.3830	0.1561	0.0919	0.0840	0.2990	0.2071
7th	0.6057	0.2260	0.0699	0.1168	0.4889	0.4190
8th	0.4545	0.3564	0.1304	0.1352	0.3193	0.1889
	Second moult period of 2 days					
9th	0.5165	0.4685	–	–	–	–
10th	0.8392	0.8890	0.4405	0.4231	0.4161	–0.0244
11th	1.1694	1.1470	0.2580	0.4761	0.6933	0.4053
12th	1.4579	1.6153	0.4683	1.0212	0.4367	–0.0316
13th	1.4903	1.6268	0.0115	0.9942	0.4961	0.4846
14th	0.8942	2.2416	0.6148	1.0008	–0.1066	–0.7214
	Third moult period of 3 days					
15th	2.1392	2.2618	–	–	–	–
16th	3.1786	2.6551	0.3933	0.9852	2.1934	1.7998
17th	4.5856	3.0899	0.4348	1.1012	3.4844	3.0496
18th	4.9988	5.2023	2.1124	1.8916	3.1072	0.9948
19th	3.9104	6.2673	1.0650	2.4218	1.4886	0.4236
20th	4.6048	6.7865	0.5192	2.4012	2.2036	1.6844
	Fourth moult period of 4 day					
21st	4.4218	9.4050	–	–	–	–
22nd	6.6618	10.3156	0.9106	1.8918	4.7700	3.8594
23rd	8.5133	14.7025	4.3869	2.6897	5.8236	1.4367
24th	10.1618	16.7038	2.0013	4.6414	5.5204	3.4708

Contd...

Table 14–Contd...

Day of Feeding	*F*	*A*	*G*	*E*	*FEA*	*R*
25th	11.5228	20.2160	3.5122	4.8626	6.6602	3.1480
26th	9.4984	21.0165	0.8005	7.4210	2.0774	1.2769
27th	11.8449	23.0850	2.0685	8.1109	3.7340	1.6655
28th	9.6535	22.8928	–0.1922	7.0014	2.6521	2.8443
29th	14.9214	23.3042	0.4114	7.0126	7.9088	7.4974
30th	11.7985	22.1109	–1.1933	8.9029	2.8956	4.0889

Daily Food Consumption and Utilization per Larva of *A. mylitta kolhapurensis* Reared on *T. arjuna*

Food Consumption (F)

A larva of *A. mylitta kolhapurensis* fed on *arjuna* consumed 0.0350 g of leaf on first day, 0.0644g on second day, 0.0871 g on third day, 0.0712 g on fourth day of feeding. Decline in food consumption was observed on the fourth day, which was the last day of the first instar. There was progressive increase in food consumption from fifth day 0.3242 g to seventh day 0.5687 g from ninth day 0.4507 g to thirteen day 1.6978 g and fifteenth day 2.1214g to nineteen day 4.9992 g of feeding. From twenty first day 4.1402 g to thirteenth day 12.6811 g food consumption was irregular. Maximum food was consumed on twenty seventh day 14.1289g.

Mean Weight of Larva (A)

Weight of freshly hatched larva of *A. mylitta kolhapurensis* was 0.0082g. On subsequent days it increased progressively till twenty-seventh day when the larva attained the maximum weight of 21.6812 g in fifth instar. Larval weight on day fifth, ninth, tenth, fifteenth, sixteenth, twenty-first, twenty second and twentyninenth day was 0.0592 g, 0.2942 g, 0.6822 g, 2.4249 g, 2.6841 g, 8.1621g, 10.2629 g and 18.6998 g respectively.

Weight Gain (G)

Initial weight gain in larva was calculated to be 0.0025 g on second day. It increased daily till twelfth day 0.5514 g and decreased on thirteenth day 0.0143 g which was the fourth day of the third

Table 15: Daily Food Consumption and Utilization per Larva of *A. mylitta kolhapurensis* reared on *Terminalia arjuna* during – Second Crop (Sep–Oct)

Unit: g.

Day of Feeding	*F*	*A*	*G*	*E*	*FEA*	*R*
1st	0.0350	0.0082	–	–	–	–
2nd	0.0644	0.0107	0.0025	0.0122	0.0522	0.0497
3rd	0.0870	0.0211	0.0104	0.0348	0.0523	0.0418
4th	0.0712	0.479	0.0268	0.0310	0.0402	0.0134
		First moult period of 1.5 days				
5th	0.3242	0.0592	–	–	–	–
6th	0.3638	0.0889	0.0297	0.0812	0.2826	0.2529
7th	0.5687	0.1742	0.0853	0.1272	0.4415	0.3562
8th	0.4281	0.2811	0.1069	0.1392	0.2889	0.1820
		Second moult period of 2 days				
9th	0.4507	0.2942	–	–	–	–
10th	0.6837	0.6822	0.3880	0.2242	0.4595	0.0715
11th	1.1218	1.1465	0.4643	0.3539	0.7679	0.3036
12th	1.2029	1.6979	0.5514	0.8992	0.3037	–0.2477
13th	1.6978	1.7122	0.0143	1.0012	0.6966	0.6823
14th	0.9842	2.3211	0.6089	0.9926	–0.0084	–0.6173
		Third moult period of 3 days				
15th	2.1214	2.4249	–	–	–	–
16th	3.6841	2.6841	0.2592	0.9442	2.7399	2.4807
17th	4.1632	3.0946	0.4105	0.9892	3.1740	2.7635
18th	4.6879	5.2014	2.1068	1.1017	3.5862	1.4794
19th	4.9992	6.6819	2.4805	2.1624	2.8368	0.3563
20th	4.1201	7.7880	0.1061	2.1013	2.0188	1.9127
		Fourth moult period of 4 day				
21st	4.1402	8.1621	–	–	–	–
22nd	6.9842	10.2629	2.1008	1.9526	5.0316	2.9308
23rd	7.6247	11.2110	0.9481	2.6342	4.9905	4.0419
24th	9.1127	13.6539	2.4429	3.1611	5.9516	3.5087

Contd...

Table 15–Contd...

Day of Feeding	*F*	*A*	*G*	*E*	*FEA*	*R*
25th	10.1028	17.8991	4.2452	4.7226	5.3802	1.1350
26th	12.2042	19.9098	2.0107	5.6893	6.5149	4.5042
27th	14.1289	21.6812	1.7714	7.9126	8.2164	4.4390
28th	10.1009	18.4268	–3.2544	5.1210	4.9799	8.2343
29th	11.3012	18.6998	0.2730	6.2097	5.0915	4.8185
30th	12.6811	19.0207	0.3209	7.2129	5.4682	5.1473

instar. In fourth instar also the weight gain decreased on the last day only *i.e.* on twentieth day 0.1061 g, Weight gain again increased from twenty fourth day 2.4429 g up to twenty-fifth day 4.2452 g Afterwards decrease in weight gain was observed till thirteenth day. Negative weight gain was observed on twenty-eighth day. Maximum weight gain was calculated on twenty-fifth day 4.2452g and minimum on twenty-eight day –3.2544 g indicating negative scope for growth during this period.

Excreta (E)

Weight of litter on second day was found to be 0.0122 g. It progressively increased on subsequent days till thirteenth day 1.0012 g, which was the fourth day of the third instar. Daily increase was observed from sixteen day 0.9442 g to twentieth day 2.1013 g followed by a decrease on twenty first day 1.9526 g which was fourth instar. In fifth instar, the weight of litter increased till twenty-seventh day 7.9125 g, which was the maximum weight of litter excreted and then decreased till twenty eighth day 5.1210 g. On the last larval day of fifth instar, 7.2129g.

Food Energy Assimilation (FEA)

FEA was 0.0522 g on second day. It decreased to 0.0402 g on fourth day of first instar. In second instar, it was 0.2826 g and 0.4415 g on sixth day and seventh day. However, it decreased to 0.2889 g on the last day of the second instar. In third instar, the FEA was maximum 0.7679 g on eleventh day. In fourth instar, it increased till nineteen day 3.5862 g, decreased on the following day 2.8368 g. FEA did not show any regular trend in fifth instar. It was 5.0316 g on twenty-second day and 5.4682 g on thirteenth day. Maximum value of FEA was observed on twenty seventh day 8.2164 g.

Respiration (R)

Respiration value was calculated to be in decrease on the last day of first instar 0.0134 g, second instar 0.1820 g and on last two days of third instar –0.6173 g. and positive highest value was also recorded on twenty-eight day 8.2393 g, twenty-fourth day 3.5087 g and twenty-sixth day 4.5042 g during fifth instar. Positive respiration was recorded on rest of the feeding days.

Daily Food Consumption and Utilization per Larva of *A. mylitta kolhapurensis* Reared on *T. tomentosa*

Food Consumption (F)

A larva of *A. mylitta kolhapurensis* fed on Asan consumed 0.0450 g of leaf on first day, 0.0745 g on second day, 0.0894 g on third day, 0.0860 g on fourth day of feeding. Decline in food consumption was observed on the fourth day, which was the last day of the first instar. There was progressive increase in food consumption from fifth day 0.2099 g to seventh day 0.5469 g from ninth day 0.3280 g to twelfth day 1.6520 g and fifteenth day 1.7668 g to eighteenth day 5.2478 g of feeding. From twenty first day 2.5404 g to thirteenth day 11.4539 g food consumption was irregular. Maximum food was consumed on twenty-fifth day 12.8451 g.

Mean Weight of Larva (A)

Weight of freshly hatched larva of *A. mylitta kolhapurensis* was 0.0082g. On subsequent days it increased progressively till thirty day when the larva attained the maximum weight of 22.3144 g in fifth instar. Larval weight on day fifth, ninth, tenth, fifteenth, sixteenth, twenty-first, twenty second and twentyninenth day was 0.0421 g, 0.4830 g, 0.7287 g, 2.0243 g, 2.7490 g, 9.5233 g, 10.1363 g and 21.3156 g respectively.

Weight Gain (G)

Initial weight gain in larva was calculated to be 0.0022 g on second day. It increased daily till thirteenth day 0.4773 g and decreased on fourteenth day 0.3241 g which was the last day of the third instar. In fourth instar also the weight gain decreased on the last day only *i.e.* on twentieth day 0.5098 g Weight gain again

Table 16: Daily Food Consumption and Utilization per Larva of *A. mylitta kolhapurensis* reared on *Terminalia tomentosa* during–First Crop (June–July)

Unit: g.

Day of Feeding	*F*	*A*	*G*	*E*	*FEA*	*R*
1st	0.0450	0.0082	–	–	–	–
2nd	0.0745	0.0104	0.0022	0.0029	0.0716	0.0694
3rd	0.0894	0.0159	0.0055	0.0103	0.0791	0.0736
4th	0.0860	0.0418	0.0259	0.0384	0.0476	0.0271
First moult period of 1.5 days						
5th	0.2099	0.0421	–	–	–	–
6th	0.3443	0.1583	0.1162	0.1606	0.1837	0.0675
7th	0.5469	0.2887	0.1304	0.2144	0.3325	0.2021
8th	0.4292	0.4360	0.1473	0.4306	–0.0014	–0.487
Second moult period of 2 days						
9th	0.3280	0.4830	–	–	–	–
10th	0.8359	0.7287	0.2457	0.2409	0.5950	0.3493
11th	1.2870	0.9670	0.2383	0.5906	0.6964	0.4581
12th	1.6520	1.4009	0.4339	0.6276	1.0244	0.5905
13th	1.1859	1.8782	0.4773	0.9317	0.2542	–0.2231
14th	1.0428	2.2023	0.3241	0.9257	0.1171	–0.2070
Third moult period of 3 days						
15th	1.7668	2.0243	–	–	–	–
16th	3.8272	2.7490	0.7247	0.7128	3.1144	2.3897
17th	5.1835	4.0534	1.3044	1.3387	3.8448	2.5404
18th	5.2478	5.0507	0.9973	1.3563	3.8915	2.8942
19th	3.7608	6.3679	1.3272	1.4570	2.3038	0.9766
20th	5.8204	6.8777	0.5098	1.1660	4.6544	4.1446
Fourth moult period of 4 day						
21st	2.5404	9.5233	–	–	–	–
22nd	5.6499	10.1363	0.6130	1.2844	4.3655	3.7525
23rd	7.6088	15.0174	4.8811	4.7940	2.8148	–2.0663
24th	9.2777	18.5123	3.4949	6.6280	2.6497	–0.8452

Contd...

Table 16–Contd...

Day of Feeding	F	A	G	E	FEA	R
25th	12.8451	21.0728	2.5605	6.2364	6.6087	4.0482
26th	8.1561	22.1539	1.0811	8.4201	–0.2640	–1.3451
27th	10.4776	22.6024	0.4485	7.0988	3.3788	2.9303
28th	8.0118	22.4275	–0.1449	6.2439	1.7679	1.9128
29th	8.4539	21.3156	–1.1119	6.9356	1.5183	2.6302
30th	11.4539	22.3134	0.9978	6.9367	4.5172	3.5194

increased from twenty-second day 0.6130 g up to twenty-fourth day 3.4949 g Afterwards decrease in weight gain was observed till twenty-ninenth day. Negative weight gain was observed on twenty-eighth day to twenty-ninenth day. Maximum weight gain was calculated on twenty-fourth day 3.4949g and minimum on twenty-ninenth day –1.1119 g indicating negative scope for growth during this period.

Excreta (E)

Weight of litter on second day was found to be 0.0029 g. It progressively increased on subsequent days till eighth day 0.4306 g, which was the last day of the second instar. Daily increase was observed from tenth day 0.2409 g to thirteenth day 0.9317 g and from seventeenth day 1.3387 g to ninetieth day 1.4570 g, followed by a decrease on twentieth day 1.1660 g which was the last day of the fourth instar. In fifth instar, the weight of litter increased till twenty-sixth day 8.4201 g, which was the maximum weight of litter excreted and then decreased till thirtieth day 6.9367 g. On the last larval day, *i.e.* thirtieth day, the weight of litter excreted was 6.9367 g.

Food Energy Assimilation (FEA)

FEA was 0.0716 g on second day. It decreased to 0.0476 g on fourth day of first instar. In second instar, it was 0.1837 g and 0.3325 g on sixth day and seventh day. However, it decreased to –0.0014 g on the last day of the second instar showing negative value. In third instar, the FEA was maximum 1.0244 g on twelfth day. In fourth instar, it increased till eighteenth day 3.8915 g, decreased on the following day 2.3038 g but again increased on the last day of the instar 4.6544 g. FEA did not show any regular trend in fifth instar. It was 4.3655 g on twenty-second day and 4.5172 g on thirteenth day. Maximum value of FEA was observed on twenty-fifth day 6.6087 g. Negative value of FEA was recorded on twenty-sixth day –0.2640 g during the fifth instar.

Respiration (R)

Respiration value was calculated to be in decrease on the last day of first instar 0.0271 g, second instar –0.4870 g and on last two days of third instar –0.2231 g and –0.2070g. Negative value was also recorded on twenty-third day –2.0663 g, twenty-fourth day –0.8452 g and twenty-sixth day –1.3451 g during fifth instar. Positive respiration was recorded on rest of the feeding days.

Daily Food Consumption and Utilization per Larva of *A. mylitta kolhapurensis* Reared on *T. tomentosa*

Food Consumption (F)

A larva of *A. mylitta kolhapurensis* fed on Asan consumed 0.0450 g of leaf on first day, 0.0845 g on second day, 0.0994 g on third day, 0.0890 g on fourth day of feeding. Decline in food consumption was observed on the fourth day, which was the last day of the first instar. There was progressive increase in food consumption from fifth day 0.2199 g to seventh day 0.5496 g from ninth day 0.3275 g to twelfth day 1.6524 g and fifteenth day 1.7667 g to eighteenth day 5.2473 g of feeding. From twenty first day 2.5444 g to thirteenth day 12.4539 g food consumption was irregular. Maximum food was consumed on twenty-fifth day 12.8457 g.

Mean Weight of Larva (A)

Weight of freshly hatched larva of *A. mylitta kolhapurensis* was 0.0082g. On subsequent days it increased progressively till thirty day when the larva attained the maximum weight of 22.8136 g in fifth instar. Larval weight on day fifth, ninth, tenth, fifteenth, sixteenth, twenty-first, twenty second and twentyninenth day was 0.0424 g, 0.4870 g, 0.7257 g, 2.0248 g, 2.7476 g, 9.5124g, 10.1654 g and 21.3159g respectively.

Weight Gain (G)

Initial weight gain in larva was calculated to be 0.0032 g on second day. It increased daily till twelfth day 0.6762 g and decreased

Table 17: Daily Food Consumption and Utilization per Larva of *A. mylitta kolhapurensis* reared on *Terminalia tomentosa* during – Second Crop (Sep–Oct)

Unit: g.

Day of Feeding	*F*	*A*	*G*	*E*	*FEA*	*R*
1st	0.0445	0.0082	–	–	–	–
2nd	0.0845	0.0114	0.0032	0.0039	0.0806	0.0774
3rd	0.0994	0.0149	0.0035	0.0103	0.0881	0.0856
4th	0.0890	0.0401	0.0252	0.0374	0.0516	0.0264
		First moult period of 1.5 days				
5th	0.2199	0.0424	–	–	–	–
6th	0.3464	0.1574	0.1150	0.1616	0.1848	0.0698
7th	0.5496	0.2877	0.1303	0.2114	0.3382	0.2079
8th	0.4270	0.4334	0.1457	0.4360	–0.0009	–0.1547
		Second moult period of 2 days				
9th	0.3275	0.4870	–	–	–	–
10th	0.8397	0.7257	0.2387	0.2400	0.5997	0.3610
11th	1.2865	0.9734	0.2477	0.5901	0.6964	0.4487
12th	1.6524	1.4019	0.4285	0.6206	1.0318	0.3556
13th	1.1853	1.8783	0.4764	0.6317	0.5536	0.0772
14th	1.0424	2.2025	0.3242	0.5257	0.5167	0.1925
		Third moult period of 3 days				
15th	1.7667	2.0248	–	–	–	–
16th	3.8265	2.7476	0.7228	0.7139	3.1126	2.3898
17th	5.1854	4.0532	1.3056	1.3378	3.8476	2.5420
18th	5.2473	5.0543	1.0011	1.3211	3.9262	2.9251
19th	3.7654	6.3654	1.3111	1.4210	2.3444	1.0333
20th	5.8256	6.8765	0.5111	1.1375	4.6881	4.1770
		Fourth moult period of 4 day				
21st	2.5544	9.5124	–	–	–	–
22nd	5.6470	10.1654	0.6530	1.8920	3.7550	3.1020
23rd	7.6365	16.0175	5.8521	5.1940	2.4425	–2.4096
24th	9.2779	18.5143	2.4968	6.6280	2.6499	–.08469

Contd...

Table 17–Contd...

Day of Feeding	*F*	*A*	*G*	*E*	*FEA*	*R*
25th	12.8457	21.0739	2.5596	6.1364	6.7093	4.1497
26th	8.1869	22.1569	1.0830	8.4204	–0.2635	–1.3465
27th	10.4787	22.6043	0.4474	8.0988	2.3799	1.9325
28th	8.0197	22.4256	–0.1787	6.2436	1.7761	1.9548
29th	9.4587	21.3159	–1.1097	6.9353	1.5234	1.6331
30th	12.4539	22.8136	1.4977	6.9358	5.5181	4.0204

on thirteenth day 0.4764 g which was the fifth day of the third instar. In fourth instar also the weight gain decreased on the third day only *i.e.* on eighteenth day 1.0011 g Weight gain again increased from ninetieth day 1.3111 g, Afterwards decrease in weight gain was observed till twenty day. Negative weight gain was observed on twenty-eighth day to twenty-ninenth day. Maximum weight gain was calculated on twenty-third day 4.8521g and minimum on twenty-eighth day –0.1787 g indicating negative scope for growth during this period.

Excreta (E)

Weight of litter on second day was found to be 0.0039 g. It progressively increased on subsequent days till eighth day 0.4360 g, which was the last day of the second instar. Daily increase was observed from tenth day 0.2400 g to thirteenth day 0.6317 g and from seventeenth day 1.3378 g to ninetieth day 1.4210 g, followed by a decrease on twentieth day 1.1375 g which was the last day of the fourth instar. In fifth instar, the weight of litter increased till twenty-sixth day 8.4204 g, which was the maximum weight of litter excreted and then decreased till thirtieth day 6.9358 g. On the last larval day, *i.e.* thirtieth day, the weight of litter excreted was 6.9358 g.

Food Energy Assimilation (FEA)

FEA was 0.0806 g on second day. It decreased to 0.0516 g on fourth day of first instar. In second instar, it was 0.1848 g and 0.3382 g on sixth day and seventh day. However, it decreased to –0.0009 g on the last day of the second instar showing negative value. In third instar, the FEA was maximum 1.0318 g on twelfth day. In fourth instar, it increased till eighteenth day 3.9262 g, decreased on the following day 2.3444 g, but again increased on the last day of the instar 4.6881 g. FEA did not show any regular trend in fifth instar. It was 3.7550 g on twenty-second day and 5.5181 g on thirteenth day. Maximum value of FEA was observed on twenty-fifth day 6.7093 g. Negative value of FEA was recorded on twenty-sixth day –0.2635g during the fifth instar.

Respiration (R)

Respiration value was calculated to be in decrease on the last day of first instar 0.0264 g, second instar –0.1547 g and on last two days of third instar 0.0772 g and 0.1925g. Negative value was also recorded on twenty-third day –2.4096 g, twenty-fourth day –0.8469 g and twenty-sixth day –1.3465 g during fifth instar. Positive respiration was recorded on rest of the feeding days.

Daily Food Consumption and Utilization per Larva of *A. mylitta kolhapurensis* Reared on *Z. jujuba*

Food Consumption (F)

A larva of *A. mylitta kolhapurensis* fed on Ber consumed 0.0425 g of leaf on first day, 0.0551 g on second day, 0.0889 g on third day, 0.1035 g on fourth day of feeding. Decline in food consumption was observed on the fifth day, which was the last day of the first instar. There was progressive increase in food consumption from sixth day 0.2030 g to eighth day 0.4589 g, from tenth day 0.3242 g to thirteen day 1.6045 g and seventeen day 3.9026 g to twenty day 6.7890 g of feeding. From twenty three day 3.3504 g to thirty eighth day 12.3539 g food consumption was irregular. Maximum food was consumed on twenty-seventh day 13.6070 g.

Mean Weight of Larva (A)

Weight of freshly hatched larva of *A. mylitta kolhapurensis* was 0.0084g. On subsequent days it increased progressively till twenty-seventh day when the larva attained the maximum weight of 8.8513 g in fifth instar. Larval weight on day fifth, ninth, tenth, fifteenth, sixteenth, twenty-first, twenty second, twentyninenth and thirty eighth day was 0.0841 g, 0.4342 g, 0.4310 g, 1.3218 g, 1.7927 g, 4.6540g, 5.8120 g, 9.4211 and 14.3910 g respectively.

Table 18: Daily Food Consumption and Utilization per Larva of *A. mylitta kolhapurensis* reared on *Ziziphus jujuba* during – First Crop (June–July)

Unit: g.

Day of Feeding	*F*	*A*	*G*	*E*	*FEA*	*R*
1st	0.0425	0.0084	–	–	–	–
2nd	0.0551	0.0132	0.0048	0.0072	0.0479	0.0431
3rd	0.0889	0.0272	0.0140	0.0110	0.0708	0.0361
4th	0.1035	0.0619	0.0347	0.0327	0.0708	0.0361
5th	0.0985	0.0840	0.0222	0.0692	0.0293	0.0071
First moult period of 2 days						
6th	0.2030	0.0849	–	–	–	–
7th	0.2840	0.2165	0.1316	0.1672	0.1168	–0.0148
8th	0.4589	0.2870	0.0705	0.2011	0.2578	0.1873
9th	0.3748	0.4342	0.1472	0.4170	–0.0422	–0.1894
Second moult period of 2 days						
10th	0.3242	0.4310	–	–	–	–
11th	0.8390	0.5635	0.1325	0.2575	0.5815	0.4490
12th	1.2805	0.7219	0.1584	0.6013	0.6792	0.9608
13th	1.6045	1.3218	0.5999	0.6267	0.9798	0.3799
14th	1.1860	1.7927	0.4709	0.5314	0.6446	–0.1163
15th	1.0840	1.9092	0.1165	0.8068	0.2772	0.1607
Third moult period of 3.5 days						
16th	1.8042	2.0449	–	–	–	–
17th	3.9096	2.8172	0.7723	0.8142	3.0884	2.3161
18th	5.1843	3.1698	0.3326	1.4110	3.7733	3.4207
19th	6.2041	3.5542	3.3844	1.6812	4.5229	4.1385
20th	6.7890	4.2430	0.6888	1.9514	4.8376	4.1488
21st	5.4210	4.6540	0.4110	1.9612	3.4598	3.0488
22nd	5.3930	5.8120	1.1580	1.3112	4.0818	2.9238
Fourth moult period of 4.5 days						
23rd	3.3504	6.7512	–	–	–	–
24th	5.6907	7.3619	0.6107	0.8118	3.8791	4.2684

Contd...

Table 18–Contd...

Day of Feeding	*F*	*A*	*G*	*E*	*FEA*	*R*
25th	7.0745	8.4645	1.1026	1.3324	5.7421	4.6395
26th	9.1801	8.6750	0.2105	1.5353	7.6448	7.4343
27th	13.6070	8.8513	0.1780	2.6642	10.9428	10.7648
28th	9.2770	8.8210	–0.0303	2.6914	6.5806	7.0053
29th	10.8040	9.4211	0.6001	3.1251	7.6789	7.0788
30th	9.3545	13.2221	3.8010	3.4108	5.8662	2.1427
31st	9.4270	12.3842	–0.8379	4.1201	5.3069	6.1448
32nd	10.3510	12.4139	0.0297	6.6814	3.6696	3.6399
33rd	11.3930	10.4215	–1.9924	7.1208	4.2722	6.2646
34th	12.3539	9.3129	–1.1086	6.1415	6.2124	7.3210
35th	9.3219	15.6748	6.3619	6.6914	2.6305	–3.7314
36th	10.1240	12.4235	–3.2513	7.1215	3.0025	6.2538
37th	10.1019	12.4841	0.0606	6.1001	4.0018	3.9412
38th	12.3539	14.3910	1.9069	7.6910	4.6629	2.7560

Weight Gain (G)

Initial weight gain in larva was calculated to be 0.0048 g on second day. It increased daily till fourth day 0.0347 g and decreased on fifth day 0.0222 g which was the last day of the first instar. In third instar also the weight gain decreased on the last day only *i.e.* on fifteenth day 0.1165 g Weight gain again increased from twenty-fourth day 0.6107 g up to twenty-fifth day 1.1026 g, Afterwards decrease in weight gain was observed till thirty-fourth day. Negative weight gain was observed on twenty-eighth day. Maximum weight gain was calculated on thirty-fifth day 6.3619g and minimum on twenty-eight day –0.0303 g indicating negative scope for growth during this period.

Excreta (E)

Weight of litter on second day was found to be 0.0072 g. It progressively increased on subsequent days till fourteenth day 0.8314 g, which was the fifth day of the third instar. Daily increase was observed from seventeenth day 0.8142 g to twenty first day 1.9612 g followed by a decrease on twenty second day 1.3112 g which was fourth instar. In fifth instar, the weight of litter increased till thirty eighth day 7.6910 g, which was the maximum weight of litter excreted.

Food Energy Assimilation (FEA)

FEA was 0.0479 g on second day. It decreased to 0.0708 g on fourth day of first instar. In second instar, it was 0.1168 g and 0.2578 g on seventh day and eighth day. However, it decreased to –0.0422 g on the last day of the second instar. In third instar, the FEA was maximum 0.9798 g on thirteenth day. In fourth instar, it increased till twentieth day 4.8376 g, decreased on the following day 3.4598 g. FEA did not show any regular trend in fifth instar. It was 10.9428 g on twenty-seventh day and 6.2124 g on thirty-fourth day. Maximum value of FEA was observed on twenty-seventh day 10.9428 g.

Respiration (R)

Respiration value was calculated to be in decrease on the last day of first instar 0.0071 g, second instar –0.1894 g and on last two days of third instar –0.1163 g. and positive highest value was also recorded on twenty-seventh day 10.7648 g, twenty-fourth day 4.2684 g and twenty-ninenth day 7.0788 g during fifth instar. Thirty fifth day –3.7314g, negative respiration, Positive respiration was recorded on rest of the feeding days.

Daily Food Consumption and Utilization per Larva of *A. mylitta kolhapurensis* reared on *Z. jujuba*

Food Consumption (F)

A larva of *A. mylitta kolhapurensis* fed on Ber consumed 0.0442 g of leaf on first day, 0.0641 g on second day, 0.0889 g on third day, 0.1142 g on fourth day of feeding. Decline in food consumption was observed on the fifth day, which was the last day of the first instar. There was progressive increase in food consumption from sixth day 0.1942 g to eighth day 0.3592 g, from tenth day 0.3069 g to thirteen day 1.4412 g and sixteenth day 1.8289 g to twenty-first day 7.1236 g of feeding. From twenty fourth day 4.1492 g to thirtynineenth day 11.3012 g food consumption was irregular. Maximum food was consumed on thirty third day 14.4992 g.

Mean Weight of Larva (A)

Weight of freshly hatched larva of *A. mylitta kolhapurensis* was 0.0083g. On subsequent days it increased progressively till thirty

Table 19: Daily Food Consumption and Utilization per Larva of *A. myiitta kolhapurensis* reared on *Ziziphus jujuba* during – Second Crop (Sep–Oct)

Unit: g.

Day of Feeding	*F*	*A*	*G*	*E*	*FEA*	*R*
1st	0.0442	0.0083	–	–	–	–
2nd	0.0641	0.0107	0.0024	0.0082	0.0559	0.0535
3rd	0.0889	0.0235	0.0128	0.0103	0.0786	0.0658
4th	0.1042	0.0549	0.0314	0.0326	0.0716	0.0502
5th	0.0989	0.0735	0.0186	0.0859	0.0130	–0.0056
First moult period of 2 days						
6th	0.1942	0.0792	–	–	–	–
7th	0.2089	0.2042	0.1650	0.1742	0.0347	–0.1303
8th	0.3592	0.2769	0.0727	0.2110	0.1482	0.0755
9th	0.3128	0.3518	0.0749	0.4270	–0.1142	–0.1891
Second moult period of 2 days						
10th	0.3069	0.3523	–	–	–	–
11th	0.6563	0.4317	0.0794	0.2570	0.3993	0.3199
12th	0.9218	0.7542	0.3225	0.6012	0.3206	–0.0019
13th	1.4412	0.8992	0.1450	0.6167	0.8245	0.6795
14th	1.1011	1.1011	0.2019	0.9208	0.1803	–0.0216
15th	1.0903	1.1028	0.0117	0.9168	0.1735	0.9051
Third moult period of 3.5 days						
16th	1.8289	1.7014	–	–	–	–
17th	3.9124	2.6918	0.9904	0.7068	3.2056	2.2152
18th	4.1824	2.9212	0.2294	1.2124	2.9700	2.7406
19th	5.0241	3.6410	0.7198	1.3435	3.6806	2.9608
20th	6.0109	3.8812	0.3402	1.3542	4.6567	4.3165
21st	7.1236	4.2410	0.2598	1.4508	5.6728	5.4130
22nd	6.0893	4.6524	0.4114	1.4612	4.6281	4.2167

Contd...

Table 19–Contd...

Day of Feeding	*F*	*A*	*G*	*E*	*FEA*	*R*
			Fourth moult period of 4.5 days			
23rd	4.1492	6.1215	–	–	–	–
24th	5.1004	8.3618	2.2403	0.8714	4.2290	1.9887
25th	8.4249	8.6817	0.3199	0.3342	7.0907	6.7708
26th	9.6811	8.9212	0.2395	1.3553	8.3258	8.0863
27th	14.4269	9.6441	0.7229	2.4508	11.9761	11.2532
28th	14.4964	9.9967	0.3526	3.1600	11.3364	10.9838
29th	10.1012	10.1221	0.1254	4.4108	5.6904	5.5650
30th	11.6212	12.6813	2.5592	6.4301	5.1911	2.6319
32nd	11.6869	13.6716	0.9903	6.5508	5.1361	4.1458
33rd	14.4992	14.4112	0.7399	5.5601	8.9991	8.1932
34th	11.1242	12.6012	–1.8100	5.5666	5.5576	7.3676
35th	11.1008	13.6114	1.0102	6.5678	4.5360	3.5228
36th	10.1109	12.0108	–1.6006	6.7201	3.3908	4.9914
37th	10.1218	11.1241	–0.8867	7.4214	2.7004	3.5871
38th	11.4210	12.6842	1.5601	6.1152	5.3058	3.7457
39th	11.3012	14.4804	1.7962	7.6892	3.6120	1.8158

three day when the larva attained the maximum weight of 14.4112 g in fifth instar. Larval weight on day fifth, ninth, tenth, fifteenth, sixteenth, twenty-first, twenty second and twentyninenth day was 0.0735 g, 0.3518 g, 0.3523g,1.1028g, 1.7014 g, 4.2410 g, 4.6524g, 9.9967 g respectively.

Weight Gain (G)

Initial weight gain in larva was calculated to be 0.0024 g on second day. It increased daily till fourth day 0.0314 g and decreased on fifth day 0.0186 g which was the fifth day of the first instar. In third instar also the weight gain decreased on the thirteenth day 0.1450g, Weight gain again increased from seventeenth day 0.9904 g up to nineteenth day 0.7198 g, Afterwards decrease in weight gain was observed till thirtyninenth day. Negative weight gain was observed on thirty fourth day. Maximum weight gain was calculated on thirty first day 2.5592g and minimum on thirty seventeenth day –0.8867 g indicating negative scope for growth during this period.

Excreta (E)

Weight of litter on second day was found to be 0.0082 g. It progressively increased on subsequent days till ninenth day 0.4270 g, which was the fourth day of the second instar. Daily increase was observed from seventeenth day 0.7068 g to twenty second day 1.4612 g, followed by a decrease on twenty-fifth day 0.8714 g which was first day of fifth instar. In fifth instar, the weight of litter irregular increased till thirty ninenth day 7.6892 g, which was the maximum weight of litter excreted

Food Energy Assimilation (FEA)

FEA was 0.0559 g on second day. It decreased to 0.0130 g on fifth day of first instar. In second instar, it was 0.0347 g and 0. 4270 g on seventh day and ninenth day. In third instar, the FEA was maximum 0.8245 g on thirteenth day. In fourth instar, it increased till twenty first day 5.6728 g, decreased on the following day 4.6281 g. FEA did not show any regular trend in fifth instar. It was 11.9761 g on twenty-eighth day and 3.6120 g on thirty-eighth day. Maximum value of FEA was observed on twenty-eighth day 11.9761 g.

Respiration (R)

Respiration value was calculated to be in decrease on the last day of first instar –0.0056 g, second instar –0.1891 g and on last day of third instar –0.9051g, and positive highest value was also recorded on twenty-eight day 11.2532 g, twenty-fifth day 1.9887 g and thirty-sixth day 3.5228 g during fifth instar. Positive respiration was recorded on rest of the feeding days.

Discussion

Saikia and Handique (1998) studied silk cocoon quality in *A. atlas* where in they reported the open type cocoon which can be spun by takli. It was observed that a single cocoon can produce, about 0.304 ± 0.007 g of silk yarn. This silk has been referred as 'Fagara' silk by Jolly *et al.* (1979).The approximate tensile strength of this spun silk was found 40 g/denier. It was also observed that the approximate tensile strength of this silk yarn is greater than the muga and tasar spun silk. The approximate tensile strength of tasar silk yarn was 22.1 g/denier (Sonwalkar, 1991). On the basis of the present study, it can be stated that it can be properly reared like other commercially important silkworms. The *A. atlas* worm will also support the rearers. In China, *A. atlas* silk is in commercial use for fine and durable silk. It also makes the fine decoration in fabric and is used for ornamental designs (Chowdhury, 1982). In India, silk of *A.atlas* is yet to be exploited for commercial use. At present no standard rearing technique is developed for *A. atlas*.The present work has added great relevance in developing rearing technique for *A.atlas*.

Rajadurai and Thangavelu (1998) studied silk cocoon quality in *A.selene* wherein they reported that the moon moth (*A. selene*) pupae undergoes summer diapause and exhibits trivoltine nature which is synchronized with the meteorological condition of the region, In Sikkim, *A.selene* is reported to pass through two generations. In summer, over wintering as a pupae (Cotes, 1881-93). Mohanraj *et al.* (1996) emphasized that *Actias callandra,* an endemic species to Andaman Islands, breeds during moist months and possibly goes for diapause as a pupa during the period when its deciduous food tree, *Lannea coromandelica* (Houtt) shed its leaves.

In the present study, observations with regard to the influence of host plant diversity on colour and morphometry of immature stages and adults, life-cycle parameters, commercial characters of cocoons are studied. The nature of colour and morphometry of various stages of moon moth have been illustrated in the present study. The male and female adults possess cascade green/pista coloured large forewing and long tailed hind wing with small ocelli on both the sides. Both the IV and V instar green coloured larvae bear yellow tubercles with spines. The amount, rate and quality of food consumed by larvae Influence the fecundity, growth rate, developmental time and survival of adults (Slansky and Scriber, 1985). All above morphological characters have been confirmed in the present form and rearing technique is designed for *A. selene.* The results of the present investigation show that the duration of developmental stages, from egg to adult was shorter for *A. selene* when reared on *T. arjuna* (58 days) than *T. tomentosa* (63 days) and *Z. mauritiana* (67days).While the difference was significant at 1 per cent level between *T. arjuna* and *Z. mauritiana,* it was not significant between *T. tomentosa* and *Z. mauritiana.* Further, the more preferred host plant *T. arjuna* was found to be superior as indicated by an increase in adult longevity and fecundity. Muiken and Brusven (1962) and McCaffery (1975) rightly pointed out that egg production in insects is influenced by the host plant selection. The commercial characters of cocoon of *A. selene* varies with the food plants *i.e.* the length, breadth, weight and shell ratio (per cent) of cocoons obtained while feeding on *T. arjuna* was significantly higher than that resulted on *Z. mauritiana* (at 5 per cent level) and on *T. tomentosa* (at 1 per cent level). The present findings are in agreement with McCaffery (1975).

The qualitative and quantitative characters of the cocoon greatly vary and much depends upon the type of food plants used (Sharma et.al 1995. Krishnaswamy et al. (1971) observed that growth and development and the economic characters of the cocoon were greatly influenced by the nutritional content of mulberry leaves in larva of *Bombyx mori* L. The pale cream, mushroom coloured and tough textured cocoons were collected and harvested from natural habitat experimental rearing.

The commercial characters of both male and female cocoons when reared on primary food plants indicate that the cocoons of *A. selene* are having sericultural importance as its counterpart *A. mylitta* cocoons. Watson (1911) in conjunction with the present observation stated that pure, white and most lustrous silk of *Actias* is valuable. But his observations are contradicting to the Deodlkar *et al.* (1969) and Mohanraj *et al.* (1996), that the papery cocoons of *Actias* are of little or no value, sericulturally, However, cocoon characterization based on the 'food plantation' needs confirmation from various regions of the world.

The papery and flimsy types of cocoons are generally obtained from secondary food plants of inferior quality. Further, Ananthakrishnan (1990) rightly pointed out that it is the host plant which plays a critical role in the determination of insect number and whole life history strategies. On the basis of our findings, it can be concluded that among the food plants. *T. arjuna* has promotive effects on the population of *A. selene* over *T. tomentosa* and *Z. mauritiana* in its natural habitat which will be helpful in conservation and utilization of moon moth at a significant level. Under laboratory conditions (Temp. 26°C– 28°C and Humidity 75–80 per cent) the rearing success of *A. selene* is about 5 per cent, 5 per cent and 2 per cent on the host plant *T. arjuna, T. tomentosa* and *T. catappa* respectively.

Antheraea mylitta Drury is potential wild silkmoth to be used in sericulture bussiness. However, at present rearing success of this species is about 35 per cent. Hence there is need to standardize the rearing technique of *A.mylitta*. Secondly, *A.mylitta* is having 44 ecoraces with different rearing potential. *A. mylitta kolhapurensis* is reported for the first time from Kolhapur, India as subspecies of *A.mylitta*. Therefore, it is very interesting to know the rearing potential of this subspecies. Hence, in the present study, rearing potential of this species have been tested and found 45 per cent which is greater than previously recorded ecoraces of *A.mylitta*. The above subspecies has been reported previously by Kavane and Sathe (2009) as variety of *A.mylitta*.

Akai *et al.* (1990) studied rearing of *A. mylitta* on artificial diet, wild silkmoth artificial diets were earlier developed for two *Antheraea* moths *A. yamamai* and *A. pernyi*. (FuKuda 1987, Higuchi, 1990), but although Indian scientists attempted to develop a diet for A. mylitta, they are not yet been successful in the production of cocoons using the diet. In a preliminary experiment, Higuchi (1990) and FuKuda (1987) prepared three types of diet and tested them on newly hatched larvae of *A. mylitta*. Three types of diet showed no repellent behavior, but showed growth rates considerably differed depending on which diet has been given. However, on only one type of diet larvae grows to maturity and spun large cocoons and on the others two types larvae did not grow to maturity. In younger stages larvae were reared in a plastic container with a small piece of diet and cotton containing distilled water. Room temperature was about 28^0c. In 4th to 5th instar, each larva was reared independently in an ice cream cup. Larval duration was not uniform and ranged from 28 to 35 days, because this species is an ecotype. Larval body weight was 31 grams. The female showed maximum 39 grams and the ratio of female: male was 10: 10. When these larvae became mature (wandering stage), they prepared two types of mounting, a mounting box containing fresh oak twigs and a plastic mounted-type nest. The larvae made cocoon easily in both types, they pupated without any trouble. Average cocoon weight of females and males was 9.77 grams (Akai *et al.*, 1990).

In contrast to the *A. yamamai* moth, *A. mylitta* after emergence copulated easily indoors, and laid 160-220 eggs per moth. In second generation Akai *et al.* (1989) successfully cultured the worms on artificial diet. However, expecting bigger insects, cocoons and cocoon filament sizes further improved diet, and the hybrid of ecotypes are essential. According to Akai *et al.* (1989) *A. mylitta* is one of the largest lepidopteran insects in its larval size, such a giant form as an experimental insect in the laboratory can be very useful in research on basic biology. With the success of the artificial diet, these giant moths may be popularized for biotechnological research in future (Akai *et al.*, 1989a, 1989b, and Akai 1990). The very thick cocoon filament is also valuable material for the development of a new type of silk fabric in the commercial silk industry.

Babu and Rao (1998) reported indoor chawki rearing tray, a new device for tasar silkworm (tropical tasar) *A. mylitta* specially designed for indoor Chawki Rearing Tray (ICRT) with hydrophonic branches as a new concept. A comparative study of indoor-cum-outdoor and field rearing of *A. mylitta* indicates that there is a significant improvement in the Effective Rate of Rearing (ERR) when the worms were reared on ICRT with hydrophonic branches up to the 3rd instar under indoor conditions and later, on outdoor plants. The overall gain of 34.75 per cent in ERR was observed with indoor chawki rearing than that in the field rearing These findings are in agreement with (Kuribayashi, 1990) over *A. yamamai* diet.

The weight of full grown larva reared in indoor chawki rearing tray was almost at par with that of the field reared lots. The average cocoon, pupal and shell weights of female have shown improvements in indoor chawki reared lots. However, in control and in the case of male cocoons difference was not significant. The shell ratio of the indoor chawki reared lots was almost at par with that of the lots reared in field. The weight of raw silk and filament length of indoor chawki reared lots were significantly more than that of the lots reared in field. Little difference was observed in the reelability and denier of indoor chawki reared lots over that of the field reared. The body weight during the indoor rearing was found to be less. There was no significant difference during chawki rearing in the growth.

According to Babu and Rao (1998) the worms after transfer from indoor chawki to outdoor rearing shown uniform mounting and larval development as compared to the field reared. This may be due to temperature with less variation maintained in the room which provided optimum condition for larval development. As reported by earlier workers, an instinct in the worms for free movement during the brushing and falling down from the tray was not found in the present Indoor Chawki Rearing on ICRT diet. The larvae were able to crawl the hydrophonic branches, which were vertically planted similar to the plants in the field. The dry weight shown significant difference in indoor chawki reared lots on ICRT with that of field reared.

Patil (1998) reported castor *Ricinus communis* as a potential new host plant for tropical tasar silkworm *A. mylitta*. The life-cycle duration in silkmoth ranged from 71 to 85 days. There was no much variation in larval development of tasar worms reared on castor leaves compared to *T. tomentosa*.

The tasar silkworms (*A.mylitta*) that were released and cultured outdoor on castor plants in crop field fed on tender leaves with no deleterious effect on their development unlike in the case of indoor culture on tender leaves of castor. In the field, the larvae of first three instars settled on the under surface of castor leaves and fed from margins inwards between mid-ribs. The brilliantly coloured IV and V instar silkworms fed also on mid-ribs from under surface, resulted in defoliation of caster leaves and escape from the predators and sharp sunlight. The larvae remained on a particular leaf till the leaf area was exhausted leaving petiole and then moved on to the next leaf. Sometimes, they also fed little portion of the anterior part of the petiole. While feeding the leaves, the worms grasp the midribs firmly and feed on the leaf lamina and midribs, they also move towards the petiole and firmly grasp it to complete the feeding of the entire leaf lamina. In bloomy variety of castor, the worms will not have the grip while moving on the trees and thus, they fall on the ground, whereas, in non-bloomy varieties and a year old castor plant, the worms will not detach from the castor plant in crop field. The full grown tasar silkworms spun dense silk cocoons with its peduncle on the leaf petiole or thin stem. The weight of tasar silk cocoon, silk shell and pupae ranged from 10 to 12 g, 1 2 to 1.5 g, 8 to 9 g respectively, which are in agreement with the cocoons produced on *T. tomontosa* (Patil, 1998).

Castor as a new and potential host of tasar silkworm, *A. mylitta* open a new vistas in tropical tasar culture and will help in boosting up of tasar silk and oil seed production benefiting the farmers to increase their farm income.

Alom *et al.* (1998) studied rearing performance of wild ecorace Modal of tasar silkworm *A.mylitta* on different food plants. The rearing performance of ecorace Modal on pruned Sal trees and defoliated Sal trees showed significant variations in number of cocoons harvested, ERR per cent, yield per dfls and cocoon weight ($p<0.01$., df=06). However, there was no difference in the shell weight and shell per cent of the cocoons ($p>0.05$, df-06). The pruning of the plants 45 days in advance of the brushing gave the highest yield when compared with all the combination of the pruned plants and this gain was found quite significant.

Defoliation of the Sal plants 46 days in advance of the brushing of the silkworm (G) has produced maximum gain in the yield per dfl. However, cocoons were of inferior quality when compared with the plants pruned 45 days in advance (C) of the brushing of the silkworm ($p<0.01$, df-06). The potential silk yield and shell per cent, however, did not show any significant variation between the pruned and defoliated plants of 45 days in advance of the brushing.

In ecorace Modal the yield per dfl with different food regimes varied between 8 cocoons on sal to 62 cocoons on asan. When the ecorace Modal was reared on different combinations of food plants, it showed significant variations in its yield and yield traits ($p<0.01$.. df=08) with maximum yield for the number of cocoons per dfl on asan (*T. tomentosa*) followed by arjun (*T. arjuna*) and within combination of these two food plants, but, it was extremely minimum on Sal (*S. robusta*).

Rearing and cocooning of tropical tasar silkworm *A. mylitta* in indoor conditions was studied by Ohja *et al.* (1994). In outdoor rearing conditions the younger larvae rest on the leaf margin and feed from the margin to the centre. The older ones attach themselves to petioles or even small twigs and feed on lamina of the leaf. After wandering stage, they start spinning of the cocoons by constructing ring of silk around a suitable twig of 0.15 mm diameter. These conditions were stimulated in the rearing boxes/trays by introducing wooden rods for freeing the margins of leaves for gripping and feeding.

The wooden rods were used by older larvae for gripping during feeding, moulting, resting and spinning. The success obtained in the present attempt at indoor rearing up to cocooning was due to the introduction of the wooden rods. Kuribayashi (1984, 1990) described Indoor rearing techniques for *A. yamamai.* The combination of 20–

22°c temperature and 40-50 per cent. RH maintained in the rearing room was necessary to minimize wilting of the diet leaves. In the rearing tray RH remained higher than that of the rearing room. RH in the tray was maximum (82 per cent). When fresh leaf diet was provided, and RH was minimum (60 per cent), then most of leaf diet was consumed (the next day). Kuribayashi (1990) reported that among the three primary food plants, *T. tomentosa* and *T. arjuna* were the most suitable for complete indoor rearing, while, the leaves of S. *robusta* were useful only for rearing fourth instar.

According to Sinha *et al.* (1998) *A mylitta, a* polyphagous silkworm feeds on a number of host plants of which *T. arjuna, T. tomentosa* and *Shorea robusta* are considered as primary food plants for commercial rearing. A feeding trial of *A. mylitta* was taken with the foliage of *T. arjuna, T. tomentosa* and *S. robusta* and food consumption, utilization and tissue growth during different larval instars have been calculated. It has been observed that an average total consumption of an individual larva was highest in *T. arjuna* (123.630 g) followed by *T. tomentosa* (105. 717 g) and lowest in S. *robusta* (94.256 g) quantitatively, difference in consumption, assimilation and respiration starts from second instar onwards depending on the types of food plant. From second instar onwards, consumption, assimilation and respiration was found to be lowest in the worms feed on *S. robusta.* Similarly, tissue growth and efficiency of conversion of ingested and digested food to body substances recorded to be lowest from third instar onwards in the worms feed on *S. robusta, T. arjuna* and *T. tomentosa* be have equally in respect of efficiency of conversion of ingested and digested food to body substances. The worms fed on *T. arjuna* were found significantly superior in respect of food consumption, assimilation, respiration, tissue growth and digestibility among all the three types of food plants, *T. tomentosa* and S. *robusta* (Sinha *et al.*, 1998).

Some salient features of Andhra local ecorace, *A.mylitta* have been reported by Purushotham Rao (1998) which are given below: The cocoon yield from the experimental plantation was significantly 800-1000 g per 1000 cocoons which was significantly higher than its yield in the forest area. The quality of the cocoons and commercial characteristics were also superior. His conclusions were based on the three crop season continued regularly for a period of three years. He concluded that isolated rearing of Andhra local ecorace with out mixing with Daba TV or other races will result in better yield. Some of the weaknesses like low fecundity; erratic emergence etc can also be controlled. In the forest areas, there is competition with exotic race Daba TV which is aggressive in nature and voracious eater. The Andhra local population thus has slowly drives to the peripheral forest region and reduced to a very limited population in the deep forest areas beyond the reach of the cultivators. There was also an interbreeding between Daba and Andhra local resulting in depletion of the cocoon qualities in the region.

Madan *et al.* (1991) and Thangavelu *et al.* (1993) made efforts to improve the rearing conditions of *A. mylitta* to reduce larval loss and increase production rate. Technological developments in recent years improved management to reduce the larval loss. Keeping this in view, efforts are being made through present investigation for a total indoor rearing from first instar larva to cocoon spinning with 45 per cent success rate.

Earlier, a few workers have tried certain improved methods of tasar rearing which include indoor rearing in polythene bags, in trays and earthen pots. Of these three methods, about 40 per cent ERR was found with polythene bag, 30 per cent with earthen pot and only 10 per cent with tray method. According to Madan *et al.* (1991) the maximum loss (50-55 per cent) of late age worms were found with polythene bag method. Similarly, Thangavelu *et al.* (1991) have reported increase in ERR of only 3.65 per cent in indoor rearing from hatching to cocoon production. 11 64 per cent increase in the ERR have been reported by Jayapraksh *et al.* (1993) when tasar worms were reared up to third instar under indoor conditions and later brought under natural plantation. The ERR (by weight) was 54 78. 68. 47 and 75. 89 with total indoor rearing of Andhra local ecorace while, the ERR of outdoor rearing was 54.81, 55.77 and 58 38 in the I, II and III crops, respectively (in the second and third crops the ERR in the total indoor rearing has increased by 12.7 per cent and 17.42 per cent, respectively). There was an increase in ERR (by number) in total indoor rearing by 7.2 per cent, 15.6 per cent and 22.8 per cent in three crops (Madan *et al.*, 1991).

In indoor rearing, it is possible to eliminate totally the loss due to pests and predators but loss due to climatic changes is possible only to certain extent and due to diseases (Griyaghay *et al.*, 1987 and Choudhuri *et al.*, 1987). It is almost same as in the outdoor rearing. A similar observation showing significant loss of worm in indoor rearing due to diseases other than pebrine was reported by a few scientists with their preliminary studies on different.species of *A. mylltta.*

According to Griyaghay *et al.* (1987) and Choudhuri *et al.* (1987) S.E.M. studies show that the filaments of outdoor reared cocoon shell are distinct from the indoor cocoons by having cross binding and a split in the filaments forming Y-shaped structures. They are also more in number and less cementing substance which is in contact to that of indoor reared cocoons. The finding that there are certain bands which are common between Daba and Andhra local ecoraces, indicates that there is a possible genetic proximity between two ecoraces.

Madan *et al.* (1998) studied development of oak tasar silkworm *A. proylei* J. in north-western Himalayas. According to them the oak tasar culture was introduced in the Sub-himalayan belt of India in early seventies with the synthesis of *Antheraea proylei* from indigenous *A. roylei* and Chinese counterpart *A. pernyi.* In spite of the fact that a vast natural oak flora is available in the country, the production of oak tasar is meagre. The major contributing factor has been the inconsistent crop performance with near to total failure of second crop. In the North-eastern region where the topography and the agro-climatic conditions are quite different from North-western region, the production of oak tasar cocoons picked up initially reaching nearly to a crore in the year 1979 in Manipur state only. However, the development of oak tasar in the North-western region was vitiated from the very beginning due to the problem of raising successful crops. In the eighties, with the screening of suitable host plant *Quercus semicarpifolia* and subsequent manipulations of rearing schedule, *a* breakthrough was achieved in the crop stability in North-western region. This gave a scope for making extension of tasar culture with the involvement of farmers.

Nayak and Rao (1998) reported larval period extended from 30 to 41 days in the Modal ecorace feeding on both the natural and artificial diets and there was no significant difference between them. No significant difference excited in survival and cocooning rate between first generation larvae feeding on both natural and artificial diets (P< 0 05), also there was no significant differer in the second generation, Cocoon weight, pupal weight, shell weight and shell ratio were reduced than the par and significantly differed ($P<0.05$) in the cocoons produced from the larvae feeding on artificial diet in both the generations. Fecundity of female moth of the larvae fed on artificial diet also reduced and differed signifies) with the parent and with that on the natural diet. Cocoon weight, shell weight, pupal weight and fecundity of silkmoth of the larvae fed on natural diet were very much like its parent. No significant difference existed in emergence, coupling and hatching of the moth when the larvae fed on both natural and artificial diets. Indoor rearing of wild Modal ecorace creates new hope to conserve this precious ecorace and its commercial cultivation for higher productivity. Cocoon, quality, silk output and fecundity can be further increased improving the quality of the artificial diet.Leaf flour the main ingredient of the artificial diet of *A. paphia* and its nutrition may directly influence the rearing result. Effect of leaf flour of different species of food plants differed much while feeding the larvae of *A. yamamai* with the artificial diet prepared out of such flour (Higuchi, 1990) Feeding the leaves of different species of food plants to the larvae of different species of the tasar silkmoth *A. mylitta* and *A. paphia* produces cocoons of different quality (Nayak, 1996). The artificial food prepared out of Sal (S. *robusta*) flour, for the larvae of Modal ecorace produce best quality of tasar cocoons (Nayak, 1998).

The leaves of *T. catappa* contain higher levels of nutrients as compared to others. Rearing performance of larvae are strongly co-related with the water and nitrogen content of the leaves (Prasad *et al.*, 2004). The foliar analysis of the *T. catappa* plant revealed that it has marginally higher moisture content (+3.26 per cent) but has a higher nitrogen content (+17 per cent) than the *T. arjuna* plant. The average larval weights (+6.76 per cent) and cocoons weight (+11.3 per cent) shows a moderate increase over arjun, while shell weight (+25 per cent) and shell ratio (+24 per cent) show significant results when reared on *T. catappa* plants. Alternative host plant *Anacardia occidentals* was used for the tropical tasar silk worm reared by Prasad *et al.* (2004) showed marginally less nutrient excepting nitrogen, which is marginally higher than the foliar nutrient content of arjun plant. The quality of leaves has direct bearing on the health, growth and survival of polyphagous tasar silk worm (Sinha *et al.*, 1986). The grainage performance of the cocoons harvested from the *T. catappa* plants seems to more or less similar when compared to that of cocoons harvested from arjun plants. The characters like pupation, pupal weight, male/female emergence, average fecundity and hatching were compared which showed that *T. catappa* is suitable for the growth, development and reproduction of tasar silkworm.

Shamitha (2007) attempted total indoor rearing of tasar silkworm *A. mylitta.* Tasar culture is a forest based industry best suited to the economy and social structure of developing countries like India. Minimum investment is the most significant feature of tasar sericulture. Tasar culture is also related to high employment and high export business. Domestication of tasar silkworm which is totally wild in nature is the major area identified to conserve, stabilize and rejuvenate the ecoraces. Most of the parameters tested in indoor rearing were found to be superior to the outdoor rearing. Larval mortality during all the three crops taken was also reduced as compared to earlier.

Recently, Kavane and Sathe (2008) studied indoor rearing of *A. mylitta* on a host plant *T.catappa* a tertiary food plant of wild silkmoth. The rearing success of *A. mylitta kolhapurensis* on *T.catappa* under laboratory conditions ($24^{\pm}10^{0}$c, 65–70 per cent R.H. and 14 hrs. photoperiod) was 45 per cent. The cocoon characters were found improved as cocoon weight, shell weight, length of shell, width of shell, shell thickness and shell ratio as compare to previous reports (Sinha *et al.*, 1986). According to Kavane and Sathe (2008) silkworms were adopted for indoor rearing technique by preparing no peduncle which is normally spun by the worms in outdoor rearing is, outstanding feature of the success of indoor rearing technique.

Chapter 7
Summary

Sericulture is related to employment, commerce and economic development of country and is a potential source of foreign exchange hence, it is increasing at a very rapid rate in India. Sericulture is practised in most of states with heterogenous agroclimatic conditions. Various mulberry cultivation methods and mulberry silkworm rearing methods have been evolved and being practised in India. However, wild sericulture in India is yet be developed for its rearing technology. Wild sericulture holds great promise for agro industry and forestry as supplementary activity. On one hand, it can help in arresting forest depletion and on other, it permits gainful utilization of vast natural wealth. The biological resources play a vital role in the economic and social development of mankind and any loss of biodiversity has serious implications on development of any nation.

Tasar sericulture needs breakthrough from the view point of either new hybrids or species/ sub species/new varieties / races which would be rearable indoor with more than 90 per cent rearing success. At present, the success of rearing of tasar silkworm is about 30-40 per cent.

The book has been divided into eight chapters.

The chapter first is devoted to general introduction which narrates national and international status of the topic.

Second chapter deals with review of literature which is upto date and relevant to the topic.

The chapter third devoted for material and methods adopted for completion of the present work. Taxonomy of wild silkmoths have been studied by collecting wild silkmoths and by making taxonomical description of them. The moths have been identified by consideration Hampson (1974), Sen and Jolly (1971) and Kavane and Sathe (2009), etc. *Attacus atlas* L, *Actias selene* Hubner and *Antheraea mylitta* sub. sp. *kolhapurensis* under laboratory conditions (28 ± 2°C) by providing their natural food plants. For *A. atlas –Angier* (*Ficus carica*) was used as food, for *A. selene – Terminalia arjuna* was used as food while, *A. mylitta* –sub. sp. *kolhapurensis–T. catappa* was used as food.

Fourth chapter is devoted for Taxonomy of wild silkmoths which includes 9 new sub species of the species *Antheraea mylitta* namely, *A. mylitta indica* sub. sp. nov, *A. mylitta jujubi* sub. sp. nov, *A. mylitta arjuni* sub. sp. nov, *A. mylitta grayi* sub. sp. nov, *A. mylitta kolhapurensis* sub. sp. nov, *A. mylitta koynei* sub. sp. nov, *A. mylitta sathei* sub. sp. nov, *A. mylitta badami* sub. sp. nov and *A. mylitta sahydricus* sub. sp. nov.

Fifth chapter embodies life history of 3 wild silkmoths. *Attacus atlas* has completed its life cycle within 35 days, *Actias selene* completed within 45 days and *Antheraea mylitta* sub. sp. *kolhapurensis* in 36 days on the food plants *F.carica, T. arjuna* and *T. catappa* respectively

The chapter sixth is devoted for rearing potential of wild silkmoths with different food plants such as *Ficus carica, Terminalia arjuna, T. tomentosa, T. catappa, Zizyphus jujuba.* The best food for *Attacus atlas* was *F. carica,* for *Actias selene* it was *T. arjuna* while, for *A.mylitta* was *T.tomentosa.*

Seventh chapter represents the summary of the book and the chapter eighth conclusion. The chapter ninenth provides bibliography referred.

Chapter 8
Conclusion

Wild Silkworms specially *Attacus atlas* L, *Actias selene* Hubner and *A. mylitta* Drury can be rearable under laboratory conditions with special care. *A. atlas, A. selene* has poor performance in laboratory conditions. However, *A. mylitta kolhapurensis* has good potential of rearing under laboratory conditions by adopting plastic boxes, G.I.trays and providing leaves of *Terminalia catappa* with 45 per cent rearing success which is better than reports published earlier. However, till there is a need to have a new hybrid varieties which will meet standard rearing potential under laboratory conditions.

References

Ahmed, S. 2003. Participation of women groups in development of tasar culture. *Indian silk,* 42(8): 19-20.

Ahsan, M. M., Jolly, M. S. and P. K. Khatri., 1974. Some preliminary studies on the various aspects of grainage of tasar silkworm (*A. mylitta* D.). Pro. *1st Int. nat. Sem. on Non mulberry silks,* Central Tasar Research Station, Ranchi, pp. 158-159.

Ahsan, M. M., Srivastava, A. K. and M.O.A. Ansari., 1979. Preservation of Bogei Cocoons at Low temperature.

Akai, H. 1998. Cocoon filament character and post cocoon technology. *3rd Int. Conf. on wild silkmoth,* Bhubaneshwar Nov. 11-15, pp. 85.

Akai, H. 1998. Global Scenario of wild silks. *Indian silk,* 37 (6&7): 18-20.

Akai, H., Asaoka, K. and P. Svacha., 1989a. Calcium crystals on Indian tasar cocoons Abs. of Kanto Branch meeting in the Japanese society at sericultural Science, 40. Japanese.

Akai, H., Soto. M., Asaoka. K., Nayak, B. K. and C. B. Jagannath Rao., 1991. *Antheraea mylitta* with newly developed artificial diet. Wild Silkmoths. 89-90. *Proc. Int. Soc. for Wild Silkmoths,* Japan, 121 – 127.

Akai, H., Suto, M. and B. K. Nayak., 1989b. Artificial diet rearing of Indian tasar silkworm, *A, mylitta*. Abs. of Kanto Branch meeting in the Japanese society at sericultural Science, 40. Japanese.

Akai, H., Suto. M., Ashoka. K., Nayak B. K and C.B. Jagannath Rao., 1990. Rearing of *A.mylitta* with newly developed artificial diets. In wild silkmoths (ed. H. Akai and M. Kiuchi), *Proc. Int. soc. for wild silkmoths*, Japan, 121 – 127.

Alam, M. O. 2006. Cocooning yard in tasar. *Indian silk*, 45(2): 23-24.

Alam, M. O., Sinha, B. R. R. P. and S.S. Sinha., 1998. Conservation and Utilization of wild tasar ecorace 'model' of *Antheraea mylitta* D. *Proc. III*[rd] *Int. Conf. on wild silkmoths*, 321-322.

Alam, M. O., Suresh M. R., Sit. S. C., Girip. Mohan Naik. B and S. Rajaram., 1991. Indoor rearing of tasar silkworm Orissa. *Indian Silk*.30 (4): 16–18.

Alam, M. O., Suresh, M.R. and S. C. Sit., 1992. Survey, collection and characterization of modal ecoraces of Indian wild silk insect *Antheraea mylitta* D.(Lepi: Satur.) *Wild silkmoth, 92*: 93-97.

Alam, M. O., Suresh, M. R. and S. C. Sit., 1993. Survey, collection and characterization of Modal ecorace of Indian wild silk insect *Antheraea mylitta* D (Lepi: Satur.). *Wild silkmoth*, 92: 93-100

Alom, M. O., Pandey, R. K., Yadav, G. S., Sinha, B. R. R. P. and S.S. Sinha., 1998. Studies on rearing performance of wild ecorace modal of tasar silkworm (*Antheraea mylitta* D.) on different food plants. *Proc. III*[rd] *Int. Conf. on wild silkmoths*, p.p. 82-86.

Ananthakrishnan, T. N. 1990. Workshop manual on insect plant interactions. Entomology Research institute, Loyola College, Madras, 2.

Ann. Rep. Central tasar research, Ranchi, India. pp. 83.

Anonymous, 1968. Indian Tasar, Directorate of commercial publicity, Government of India, New Delhi, pp.32.

Arora, G. S. and J. J. Gupta., 1979. Non Mulberry Silkmoths. *Mem. of Zool. sur. India*, 16 (1): 6-35.

Babu, S and Purushotham Rao. 1998. Indoor chawki rearing trey- A new device for tasar silkworm *A.mylitta* D. *Proc. III*[rd] *International conference of wild silkmoths*. Pp. 87-91.

Bari, A. and A. Purushotham Rao., 2002. Physico chemical studies on *Antheraea mylitta* Drury cocoon shell. (Satumidae-lepidoptera). *Uttar Pradesh J. Zool.* vol. 22(1): 45-51.

Bari, A. and A. Purushotham Rao., 2003. Physical and Commercial Characteristics of certain Indian wild Silkmoth *Antheraea mylitta* Drury (Saturniidae–lepideptera). *Uttar Pradesh J. Zool.* 23(1): 39-51.

Barlow, H. S. 1982. An introduction to the moths of south east Asia. The Malayan nature society, Kuala Lumpur, 305.

Baruh, A., Gorwami, M. C and M. V. Samson., 1989. Consumption and utilization of food in different instars of muga silkworm *A.assama* Ww. *Proc. Indian Acad. Seri.* (98): 99 – 103.

Beeson, C. F. 1941. The ecology and control of the forest insects of India and the neighboring countries, Pub. Mahendra Pal Singh. Dehra Dun, 695-696

Behera, B. 1997. Biometry and typology of tropical tasar cocoon of *A. mylitta* Drury (Lepidoptera: Saturniidae). M. Phil. Dissertation, Utkal University. pp. 1-16, 17-51, 55-86.

Bhattacharya A and R. S. Teotia., 1998. Conservation strategies of wild silk moths in the North Eastern region of India. *Proc. III*rd pp. 311-313.

Chakravorty, K. 1997. Forestry and Sustainable development. *Everyman's Science*, 31(5): 148-152.

Chakrovoty, R. and B. C. Das., 1979. Studies on disease of tasar food plant and their control measure. Ann. rep. Central Tasar Research, Ranchi, India. 139-141

Chandra, H. 1997. Studies on aero–phyllo microflora of oak (Q. serrata Thumb) in Manipur with special reference to certain bacterial diseases of Oak Tasar Silkworm (*A. proylei* Jolly). In Ph.D. Thesis: pp. 1-2.

Chaoba singh, K. and Suryanarayanna., 2005. Wild silkmoth wealth of India. *Proc.Advance in tropical sericulture.* (edit. Dandin and Gupta) Pp.419 – 421.

Choudhuri, C. C., Dubey, O. P. Sen Gupta, K. and S. S. Sinha., 1987 b. Indoor rearing of tasar silkworm upto 144 hours. Annual reports. CTR and TI Ranchi.

Choudhuri, C. C., Dubey; O. P., Sinha, S. S. and K. Sen Gupta., 1987a. Studies on the different techniques of young stage silkworm rearing, Annual report CTR and TI Ranchi.

Chowdhury, S. N. 1982. Muga silk industry. Directorate of sericulture, Govt. of Assam. 8–11.

Cotes, E. C. 1891-1893. The wild silk insects of India. Indian Mus.Notes. Calcutta. 2, 69-89.

Cotes, E. C. 1942. Silkworms in India. Indian Mus. Notes, Calcutta.1(3). 129-173.

Cotes, E. C. and Swinhee., 1889. Catalogue of the moths of India. P.P. 288.

Cotes, E. C., 1889. Silkworm in India. Indian Museum Notes, 1(3): 157-162.

Danilevskii, A. S. 1965. Photoperiodism and seasonal development of insect. Oliver and Boyd, London, pp. 1-123.

Dash, A. K. and B. K. Nayak., 1990. Studies on some oviposition behaviours of Godamodal ecorace of Indian wild tasar silkmoth *A. paphia* L. (lepidoptera: Saturniidae). *Proc. Ist Int. conf. on wild silkmoth*, China, No. 118, pp.5.

Dash, A. K. and B. K. Nayak., 1991. Conservation of Goda Model ecoraces of Indian wild tasar silk insect, *Antheraea paphia* Linn. (Lepi.: Saturniidae). *Sericologia,* 31 (1), 209-212.

Dash, A. K. and B. K. Nayak., 1992. Studies on some oviposition behaviours of Godamodal ecoraces of Indian wild tasar silkmoth *Antheraea mylitta* linn. (Lepidoptera – Saturniidae). *Wild Silkmoths, 91.* 78-82.

Dawson, R. W. 1993. The problem of voltinism and dormancy in the polyphemus moth. *J. Exp. Zool*, 59: 87-131.

Deodikar, G. B., Kshirsagar, K. K. and, I. A. Kamate., 1969. Chromosome number in *Actias selene* Hubner a wild silkworm with reelable cocoons. *Ind. J. Genet plant Breeding*, 29, 126-130.

Dhal, A. 1999. Impact of a biotic factors on the development of tasar cocoon of *A. mylitta* Drury (Lepidoptera–Saturnidae). M. Phil. Dissertation, Utkal University, pp. 1-6, 7-19, 42-52.

Dikshit, B. K. 2007. Similipal biosphere reserve, A natural habitat of modal ecoraces of *Antheraea mylitta* Drury. *Indian silk*, 45(9): 11-14.

Draudt, M. 1929 (1940). Saturnidae (American) In A. Seitz. The macro Lepidoptera of the world, 6, 713-827.

Dutta, B. M. 2006. The north–east, eco-friendly approach in sericulture. *Indian silk*, 45(4), 24-26.

Dutta, R. K. 1988. Indian Sericulture: Past, present and future. I[st] cong. Tropical sericulture practices, 18-23. Feb. pp. 217-240.

Essig, E. O. 1941. College Entomology, Satish book enterprise, Agra, India.491-492

Fabricus, J. C. 1793. Entomologia Systematica emendate et. aucta-Hafniae, Impensis, C. G. profit. fil. et. soc.3(2), 350.

FAO, 1979. Manuals on sericulture. Central silk board, Bangalore. Pp1-178.

Fletcher, T. B. 1914. Some South Indian Insects, Dehra Dun, 209.

Fukuda. T. 1960. Artificial diet rearing of silkworm. Seric.*Sci. Japan*, 29(1).

Fukuda. T. 1987. Artificial diets for *Bombyx mori* and *Antheraea yamamai.* Japan Sericulture News Press.

Ghosh, S. S. and D. Sengupta., 1980. Studies on reeling of emerged tasar cocoons. Ann. Rep. Central tasar research, Ranchi. India. PP. 107-108.

Ghosh, S. S. and P.S. Lamba., 1976. Studies on preservation of cooked Daba cocoons for reeling. Ann. Rep. Central tarar research, Ranchi, India. pp. 144-147.

Ghosh, S. S., Sreenivasa, D., Sengupta and K. Thangavelu., 2001. Application of lac dye on tasar silk. *Indian silk*, 40(4), 27-29.

Goel, A. K., Bramhachari, B.V., Thandapani, M and K. Thangavelu., 1993. Socioeconomic study of tasar culture. *Indian silk*, 31(12): 38-42.

Gohain, R. and R. Baruah., 1983. Effect of temperature and humidity on development, survival and oviposition in laboratory population of eri worm, *Philosamia ricini, Arch, Internat. Physiol.biochem*, 91, 83-93.

Gopalchar, A. R. S. 1978. Sericulture industry in India: potentials and prospects. Central silk board, Bombay, pp.1.

Griyaghey, U. P., Sengupta, K., Kumar. P., Rammurthi, Sinha A.K. and U. S. P. Sinha., 1987. Studies on the effectiveness of Bengard (r) in controlling the microsporidian disease of *A. mylitta* D. *Sericologia.* 27 (3): Pp. 533-540.

Guene'e, A. 1852-4. In Boisduval, J. A. and Guenee, A. Histoire naturelle des insects species general des. Lepidopteras. 5 Noctuelides. I. xcvi + 407, pp. 6, Noctuelides. II, 444. pp. 7. Noctuelides, 111, 442, pp.8. Deltoides et Pyralites, pp. 448, Paris.

Gupta, R., A. K. Singh., K. K. Chatterjee., H. Chandra and D. Chakraorty., 2008. Tasar culture showed them the way. *Indian silk*, 46(10).20-21.

Hampson, G. F. 1894. The fauna of British India, including Ceylon and Burma, moth II. XXXII + pp.609. London.

Hampson, G. F. Tailor and Francis, 1991. Revised edition, Fauna of British India, Moths, London.

Higuchi, Y. 1990. Rearing artificial diet Ten Sun. *Technology* (Eds. H. Akai and S. Kuribayashi) Science House, ToKyo.124-135.

Imms, A. D. 1957. A text book of entomology. Chapman and Hall (London).

Islam, M. 1990. The flora of majuli Assam. Pub. B. Singh and M. singh, Dehardun. 161–162.

Jadhav, A. D., Kalantri L. B., Hajare., Salunkhe T.N., Kirsur D.K., Balsaraf M.V., and T.V. Sathe., 2007. Future prospects of eri culture in the state of Maharashtra. *Serichal*. Vratza, Bulgaria. 355- 359.

Jaipal, Reddy, K., Jayaprakash P., Yadav, G. S., Seshababu V. and B. R. R. P. Sinha., 1998. Utilisation of Andhra and Bhandara local ecoraces (Wild tasar silk moths) for the prospective production of tasar cocoons. Proc. IIIrd Int. Conf. on wild silkmoth, p.p. 348-355.

Jayaprakash, P. 1995. Bioecology of tasar silkworm Andhra local *A.mylitta* from forest areas of Warangal. Ph.D. Thesis to Kakatia University, Warangal. India.

Jayaprakash, P., Naidu, W. D and M. Vijay Kumar., 1993. Indoor rearing of tasar silkworm *A. mylitta,* a new technology. Proc. of the workshop on tasar culture, Warangal, A. P. 19-24

Jolly, M. S. 1972. A new technique of tasar silkworm rearing. *Indian silk.* 11, 5-8.

Jolly, M. S. 1973. Discovery of new field of tasar on Oak and its impact on silk production, export and employment. Central tasar research station. Ranchi. Pp.3.

Jolly, M. S. 1974. New Dimension of Tasar industry in India. Souvenir. Central silk board, Bombay XII, 83-86.

Jolly, M. S. 1976. Package and practices for tropical tasar culture, Central tasar research, Ranchi, India, pp. 32.

Jolly, M. S. Chowdhary, S. N. and S. K. Sen., 1977, Non Mulberry Sericulture in India, Central Silk board, Bombay, 28-73.

Jolly, M. S., Ahsan, M. M. and S.R. Vishakarma., 1973. Rearing performance of tasar silk worm with interchange of food plants. Ann. Rep., Central Tasar Research, Ranchi, India, pp, 67-69.

Jolly, M. S., Chaturvedi, S. N and S.A. Prasad., 1968. Survey of tasar crop in India. *Indian J. Seric,* 1, Pp. 50–58.

Jolly, M. S., Chowdhary, S. N. and S. K. Sen., 1975. Non–Mulberry Sericulture in India. Central Silk Board, Bombay, pp. 1-89.

Jolly, M. S., Narasimhanna, M. N and V. N. Badaiyar., 1969. Almond body mutant in *A. mylitta,* its origin and pattern of inheritance. *Genetica,* 40, 421-426.

Jolly, M. S., Narasimhanna. M. N., Sinha. S. S and S. K. Sen., 1969. Interspecific hybridization in Antheraea. *Indian J. Hered.* 1. 45-48.

Jolly, M. S., Sen, S. K. and M. G. Das., 1975. Tasar Culture: a potential forest based industry Doc, 2d FAO / IUFRO World technical consultation on forest disease 25 pests. p.p. 14.

Jolly, M. S., Sen, S. K., Sonwalkar, T. N. and G.K. Prasad., 1987. FAO Agricultural services Bulletin (1987). Mannuals on sericulture, Non mulberry silks, Vol.4 and Mulberry cultivation, Vol. 1. Central silk Board, Bangalore, India 2-150 and 4: 1-178.

Jolly, M. S., Sonwalkar T., Sen S. K. and G. K. Prasad., 1979. Non mulberry silks. FAO manual on sericulture. Food and Agriculture organization of the united nations, Rome.

Jolly, M. S., Waratkar, V. B. and S. Prasad., 1967. A new mutant of tasar silk worm *A. mylitta* D. *Ind. J. Ent.*, 29: 389-390.

Jordan, K. 1911. Descriptions of New Saturniidae. *Novit zool.*18.129–134.

Kakati, L. N. 2006. Vanya sericulture in Nagaland problems-prospects. *Indian silk*, 45(3).18-22.

Kalantri, L. B., Jadhav, A. D., Hajare, T. N., Salunkhe, D. Y., Undale, J.P., Balsarat, A.V and T.V.Sathe. 2007. Progress and prospects of sericulture industry in Maharashtra. India. *Serichal*, pp.360–368.

Kapila, M. L. 1990. Tasar silkworm rearing can help environmental improvement and social forestry in background area. *Indian silk*, 18 (12): 6-9.

Kato, Y. and S. Sakate., 1981. Studies on summer diapause of *A. yamamai* (Lepiodoptera – Saturniidae) III. Influence of photoperiod on larval stages. *Appl. Ent.*, 16 (4): 499-500.

Kavane, R. P. and Sathe, T. V. 2007. Indoor rearing of tasar silkworm *Antheraea mylitta* D. on *Terminalia catappa* Cooke. Biotechnological approaches in Entomology. Manglam Publications, New Delhi. 178-183.

Kavane, R. P. and Sathe, T. V. 2009. On a new variety (kolhapurensis) of *Antheraea mylitta* from India. *Biological forum – An international journal*.1(2).45-47

Kavane, R. P., S. H. Thite., A. D. Jadhav and T. V. Sathe, 2004. Studies on natural enemies of tasar silkworm *Antheraea mylitta* D. From Satara district of Maharashtra. *Bull. Bio. Sci.* (1), 77- 80.

Kehimkar, I. D. 1997. An introduction to moths NCSTS. Hornbill series, Mumbai – Dec, 1999, pp, 379.

Kimothi, R. C. 2004. Oak tasar in Uttaranchal: towards improving productivity norms, *Indian silk*, 42(12), 20-23.

Kirsur, M. V. and J.V. Krishna Rao., 2003. Wild silks of India and their brand identity. *Indian silk*, 42(4): 23-26.

Kishore, Ram., Sharma B.P., S. K. Sharan. and B.R.R.P. Sinha., 2002. IPM approach to optimize tasar silkworm cocoon production. Advance in Indian sericulture research (edit. Dandin and Gupta).Pp.402–405.

Krishnaswami, S., Kumar Raj, S., Vijayaraghavan, K. and K. Kasiviswanathan., 1971. Silkworm feeding trials for evaluating the quality of mulberry leaves as influence by variety. Species and nitrogen fertilization. *Indian j. Seric.* 9(1): 79 – 89.

Kuman, P. A. and S.S. Sinha., 1974. Effect of photoperiod and temperature on pupal diapause in *A.proyeli. Proc. 1st Int.Sem. Non-mulberry silks*, pp.135-142

Kumar, A. 1993. Rearing of tropical tasar silkworm on Sal flora. *Indian silk*, 31(12), 52-55.

Kumar, R., S. B. Zeya., A. K. Srivastava., Gargi and N. Suryanarayan., 2009. Tasar flora status and potential utilization. *Indian silk*, 47(12): 20-29.

Kumarraj, S. 1968. Further studies on double cocoons of *Bombyx mori*L. *Ind. J. seric.*7(1): 70-71.

Kuribayashi, S. 1990. An improvement on the indoor rearing of the Japanese oak silkworm *A. yamamai. Proc. Int. Conf. On Wild Silkmoths*. Shenyank. China.

Lefory, H. M. and C. C. Ghosh., 1912. Eri Silk. Mem. Dept. Agri India. *Ent.*4(1): 130

Linnaeus, C. 1758. System nature Edn.10, Regnum Ahimale, 1: 824. Holmaie.

Madan, M. Bhat and Sinha, B.R.R.P., 1998. Development of Oak Tasar silkworm, *Antheraea proylei* J. in north western Himalayas. *Proc. IIIrd Int. Conf. on wild silkmoth*, 259-361.

Mani, M. S. 1993. Modern classification of insects. Pub. Satish Book Enterprise, Moti katra, Agra, India, pp.1-131.

Mansing, A and Smallman, B, N., 1967. Effect of photoperiod on the incidence of physiology of diapause in the two saturniids. *J.insect physiology.*, 13: 1147-1162.

Mathur, S K. and Shukla, R. M.1997.Tasar silkworm rearing technology silkworm disease and pest management *Current technology seminar on non-mulberry Sericulture*, conducted by Central Tasar Research and Training Institute, Ranchi.

Mathur, S.K., Singh, B.M., Sinha, A.K.and Sinha, B.R.R.P., 1998. Integrated package for rearing of *A.mylitta* D. *Indian silk*, 37(2): 15-18.

McCaffery, A. R. 1975. Indian insect life. Today, tomorrow, printers and pub. New Delhi

Mira Madan, Neeru Sailaja and Padma Vasudevan.1991. Polythene bag methods. A new technique for rearing Tasar Silkworm till the third instar. *Indian silk*, 37(2): 15-18.

Mishra, S.N., 2002. Sericulture industry in Indonesia, *Indian silk*, 41 (7): 24-25.

Mitra, Gautam and M.A.Moon., 2009. Raily –an important eco-race of *A. mylitta* Drury. *Indian silk*.47 (10): 20-21.

Mohan Rao, K., and B. Satpathy, 2003. Modal –A unique tasar ecorace. *Indian silk*, 42(5): 15-17.

Mohan Rao, K., B. Satpathy, H. C. Mohapatra and V.Kulshrestra, 2004. Grainage and rearing behaviour of modal in ex-situ conditions. *Indian silk*, 43(2): 13-14.

Mohan Rao, K., S. K. Patnaik, S. Satpathy, M. Dewanji and R. K. Sinha, 2008. Forest protection force and tasar development. *Indian silk*, 47(5): 18-20.

Mohanraj, P., Veenakumari, K and Peigler, R.S., 1996. The host plant and preimaginal stages of *Actias callandra* (Saturniidae) from the Andaman islands, India. *J.Res.Lepid.*, 32, 16-25.

Mohanty, P. K., 2002. Ecofriendly Indian wild silk culture: an approach for sustainable development, campus diversity, Initiative (CDI) monograph series–3 (Food Foundation Assisted Project), Utkal University, Bhubanswar.

Mohanty, P. K., Patro, K. B. G. and Nayak, B. K., 1997b. Conservation of tasar biodiversity through some improved methods; a need for better productivity tribal economy and sustainable use of forest. *Proc. 17th. Int. Sericultural cong.*, April 22-26, Londrina, Brazil.

Mohanty, P.K and Mohanty, S.B., 1997. Conservation of tasar biodiversity and sustainable development. *Bull. Ind. Acad. Seri.*, 1(1): 14-20.

Mohanty, P.K. 1991. Studies on some aspects of biology and ecology of *A. paphia* Linn. (lepidoptera: Saturriidae). *Proc. 78th Ind. Sci. Cong. Indore*, pt. III, pp. 93.

Mohanty, P.K. 1994. Tasar Silk: The pride of Orissa. Orissa Review, 51 (5): 1-3.

Mohanty, P.K. 1998. Tropical tasar culture in India. DPH, New Delhi, pp. 1-14, 15-39, 40-93, and 116-129.

Mohanty, P.K. and Behera, B., 1998. Biotype of tasar cocoons of *A. mylitta* Drury (Lepidoptera – Saturniidae), *Proc.85th Ind.Sci. Cong., Hyderabad*, pt.III, pp.18-19.

Mohanty, P.K. and Behera, B., 1998. Tasar culture and eco-friendly attributes. *Zoo print*, 12(5): 20-21.

Mohanty, P.K. and Behera, M.K., 1995. Some observations on hammock formation in tasar silk insect *A. paphia* Linn. (Lepidoptera – Saturniidae), *Ad. Bios*. 15(1): 1-4.

Mohanty, P.K. and Behera, M. K., 1995.Tasar cocoon and tribal belief. *Indian silk*, 34 (4): 44.

Mohanty, P.K. and Behera, M.K. 1996. Moratility of tasar silkworm *A. paphia* L. (Lepidoptera: Saturnidae) due to pebrine (*Nosema sp.*) infection. *Environment and ecology*, 14(2): 358-360.

Mohanty, P.K. and Behera, M. K., 1997. Emergence and mortality behaviour of tropical wild tasar moth *A. paphia* Linn. *Proc.84th Ind. Sci. Cong.*, New Delhi, pp.18.

Mohanty, P.K., 1994. Biology and some ecological aspects of silkmoth *A. paphia* Linn (Lepidoptera: Saturniidae). Ph.D thesis of Sambalpur University, Sambalpur, Orissa.

Mohanty, P.K., 1995. Wild tasar moth and its conservation. *Zoo print*, 10(10): 1-3.

Mohanty, P.K., 1997. Species diversity in tasar silk moth and its future Scenario. *Zoo print*, 12 (2): 5-6.

Mohanty, P.K., 2000. Conservation of Indian tropical wild tasar silkmoths and its farming prospects. *Proc. 2nd Int. workshop of conservation and utilization of commercial insects*, pp, 85-118.

Mohanty, P.K., 2001. Entinction (Ed. Bell, C.E.). Encyclopaedia of the world's Zoos, Fitzroy Dearbern publishers, Chicago and London, Vol.I (A-F), pp.457-460.

Mohanty, P.K., Das, M.C. Mishra, C.S.K. and Nayak, B. K. 1992. Seasonal Impact on life span of Tasar moth. *A. mylitta. Proc. 79th Ind. Sci. cong.*, Baroda, Pt. III pp-65.

Mohanty, P.K., Dhal A. and Pandit, J.K. 1999. Tasar culture: Spider for biocontrol. *Indian silk*, 37(10): 15-16.

Mohanty, P.K., Dhal, A. Bwehera B. and Pandit S.K. 1998. Sal Borer: a menance to sal forest and tasar culture. *Indian silk* 37(5): 24-26.

Mohanty, P.K., Dhal, A., 1997. Egg laying behaviour and certified egg production in indian wild tasar silkmoth *A.mylitta* D. *Zoo print*, 12(10): 11-12.

Mohanty, P.K., Dhal, A., Behera, B. and Pandit, J.K., 1998 Conservation of modal ecoraces of Indian wild tasar silkmoth *A. paphia* Linn. and farming prospect, Souvenir, *Proc., natl., Sem., Bakewar* (*Etawah*), U.P., feb. 06-08, 54-55.

Mohanty, P.K., Dhal, A., Behera, B.and Pandit, J.K., 1998. Conservation of tasar biodiversity through some improved methods: a need for better productivity, tribal economy and sustainable use of the forest. *Int. J. Env. And Ecoplan.*, 1 (1 and 2): 57-65.

Mohanty, P.K., Dhal, A., Mohaputra, P.K., and Mohanty, B. 2000. Impact of abiotic factors on development of tasar cocoons of *A. mylitta* Drury (Lepidoptera – Saturniidae), *Proc.87th. Int.Sci.Cong.*, Pune, pt.III, pp.55.

Mohapatra, P.K. 1999. Larval development of tasar silkmoth *A. paphia* Linn. (Lepideptera–Saturnidae) M. Phil. Dissertation, Utkal University, pp. 1-10, 13-83.

Moon, M. A and Gautam Mitra., 2007. Tensile properties of tasar reeled yarn. *Indian silk*.46 (1): 20-22.

Moore, F., 1859. Synopsis of the Known Asiatic species of silk. Producing moths, with description of some new species from India. *Proc. Zool. Soc.* Lon. 27. 237-270.

Moore, F., 1872. Description of New Lepidoptera. *Proc. Zool. Soc.* Lon. 555-583, pls.32-34.

Moore, F., 1877. The Lepidoptera fauna of the Andaman and Nicobar islands, *Proc.Zool.Soc.*Lon. 580-632.

Nadler, L. 1984. The handbook of HRD, New York: John Wiley and sons.

Nagaraja, G.M and Nagaraju J., 1995. Genomic fingerprinting of silkworm, *Bombyx mori* using random arbitrary primers. *Electrophoresis*. 16: 1639 – 1642.

Nagaraju, J and Jolly M. S., 1986. Interspecific hybrids of *Antheraea pernyi* and *A. roylei* – A cytogenetic reassessment. *Theoretical and Applied genetics*. 72: 269 – 273.

Nagaraju, J. 1998. Silk yield attributes correlations and complexities, in silkworm breeding (Ed. G. S. Reedy) Oxford and IBH publishing, New Delhi.

Nagaraju, J. and Damoder Reddy. K., 1998. Non mulberry silkworms some biotechnological perspectives. *Indian silk*, 37 (6 and 7). 43 – 47.

Nagaraju, J. Sharma A, Sethuraman B. N, Rao G.V, and Singh L, 1995. DNA fingerprinting in silkworm *Bombyx mori* using banded krait minar satellite DNA derived probe. *Electrophoresis* 16: 1639 – 1642.

Narain, Rai., B. C. Prasad, Dinesh kumar., Shobha Beck., J.Tirkey and G.C.Roy, 2002. Seed organization in tasar – A new concept. *Indian silk*, 41(7): 15-18.

Narasimhanna, M. N. and Jolly. M.S. 1969. Developmental morphology of tasar silkworm *Antheraea mylitta D* (Lep: Saturniidae). *Indian J. Seric.* 8(l): 49-52.

Nassig, W.A. and Peigler, R.S.1984. The life history of *Actias maenas* (Saturniidae). *J. Lepid. Soc*, 38, 114-123.

Nayak, B K, Dash. A. K and Dash, M. C.1988. Production of seed cocoons of *Antheraea paphia* Linn (Lepi: Satur.) by rearing on Sal plants (S. *robusta*). *Proc Int. Cong, on Tropical Sen.* Bangalore.

Nayak, B. K., Dash, A.K. and Patro K.B.J., 2000. Biodiversity conservation of wild tasar silkmoth *Antheraea paphia* Linn. of Simipal bioreserve and strategy for its economic utilization. *Int. J. Wild silkmoth and Silk*, 5: 367-370.

Nayak, B. K., Pradhan, K.C. and Satpathy, B.N., 1985. Studies on Asan (*Terminalia tomentosa*) leaf consumption by successive instars and during the entire larval period of *A.mylitta* Drury. *Sericologia*, 25 (1): 26-31.

Nayak, B.K and Dash, M.C., 1997. Save our tasar: an appeal. *Bull. Ind.Acad.Seri*, 1(1): 52-59.

Nayak, B.K. and Dash, M.C. 1991 Environmental regulation of voltinism in *A. mylitta* Drury (Lepidoptera: Saturniidae), the Indian Tasar silk insect, *Sericologia*, 31(3): 479-486.

Nayak, B.K. and Miller, E. William. 1998. The body size and altitude correlation in Lepidoptera: New findings from tropical *Antheraea* Silkmoth (Saturnildae). *Bull. Ind. Acad. Seri.*, 2(2): 16(20).

Nayak, B.K. Gupta, M.L. Dash, A.K. and Guru, B.C. 1986. Sex relation in Chupi Cocoon of Tasar silk insect *A. mylitta* Drury. (Lepidoptera–Saturniidae). *Science and culture*, 52(10): 349-350.

Nayak, B.K., and Jagannatha Rao, C.B., 1998. Domestication of the Wild Tasar silk moth by natural and artificial diets. *Proc. III*[rd] *Int. Conf. on wild silkmoth*, p.p. 75-77

Nayak, B.K., Dash, A.K. Dash. P.K., Sasmal.A.K. and Satpathy, B.N., 1986. Sex association in double cocoons of tasar silk insect *A. mylitta* D.(Lepidoptera – Saturniidae). *Sericololia*, 26(3): 285-290.

Nayak, B.K., Dash, A.K., Guru, B.C.and Satpathy, B.N., 1988. Sex association in bishellate cocoons of tasar silk *A. mylitta* D. (Lepidoptera – Saturniidae), *Entomon.*, 13 (1): 91-93.

Nayak, Braja Kishore, Dash Amulya Kumar and Patro K.B.G., 1998. Biodiversity Conservation of Wild Tasar silkmoth *Antheraea paphia* Linn. of Simplipal biosphere reserve and strategy for its economic utilisation. *Proc. III*[rd] *Int. Conf. on wild silkmoth*, p.p. 367- 370.

Nayak. B. K., Dash, A.K., Mishra, C. S. K., NahaK, U. K., Dash, M C. and Prabhakar, 0. R. 1994. Innovation of technology for commercial rearing of Indian wild tasar silk Insect, Goda modal ecorace *of Antheraea paphia* Linn. (Lepidoptera: Saturniidae). *Int. J. Wild Silkmoth and Silk* 1: 75–79.

Nayar (Jr) B.K., and Guru B.C., 1998. Studies on Cocoon biometry of the bivoltine tasar silkmoth, *Antheraea mylitta* Drury (Lepidoptera: Saturniidae) regred at Kuliana area of Mayurbhanj District. *Orissa. proc.* p.p. 281-283.

Ojha, N.G., Sinha S.S., Singh M.K and Sharan S.K.1994. Rearing and cocooning of tropical tasar silkworm *A. mylitta* in indoor condition. *Int. J. Wild Silkworm and Silk.* 1 (2). 257 – 260.

Pandey. R. K, . 1995. Indoor oak tasar silkworm rearing in North-East. *Indian Silk.* 33(10): 29-31

Pandit, J.K., 1998. Seasonal impact on cocoon characters of tropical tasar silkmoth *A. mylitta* Drury (Lepideptera–Saturnidae). M. Phil. Dissertation, Utkal University, p.p. 1-4, 7-28.

Pareek, U and Rao, T.V. 1975. HRD System in Larson and Toubro. Ahmedabad: Indian Institute of Management.

Pareek, U., Aahad, M. O., S. Ramnarayan and T. V. Rao (Eds.). 2002..Human resource development in Asia: Trend and challenges, New Delhi: Oxford and IBH.

Patil, G. M. 1991.Biology and anatomy of tasar Uzi fly (Diptera – Tachinidae) on tasar silkworm under Bangalore conditions Ph. D Thesis. Karnataka University, Dharwad (India). 230.

Patil, G. M. 1996.Domesticated tasar silkworm strain on castor, UAS News Letter, University of Agricultural Science. Dharwad, 12(1-3)5.

Patil, G. M. and Savanurmath, C. J.1989. Can tropical tasar silkworm *Antheraea paphia* (Linn) be reared indoors? *Entomon.* 14(3 and 4), 217-225.

Patil, G.M., 1998, Caster *Ricinus Communis* L.–A Potential new host of tropical tasar Silkworm, *A. mylitta* Drury (Lepidoptera: Saturnidae). *Proc. III*[rd] *Int. Conf. on wild silkmoth*, p.p. 51-53.

Periasamy, K. 1986. Problems and prospects of sericulture. lectures of sericulture, Bangalore, pp. 98-107.

Potter, A. L. 1941. The Chinese moon moth. *Actias selene.* Hong kong. *Naturalist.* 10, 167 – 172.

Prasad, B.C., Negi B.B.S., Singh. G. S., Saxena S. K. and Sinha B.R.R.P. 2005. Tasar silkworm brushing bag. *Indian silk.* 43 (9) 20 – 32.

Prasad, B.C., Rath. S. S., Nagaendra S and Sinha B.R.R.P.2004. Cashew – another food plant for tropical tasar silkworm. *Indian silk*, 43(1), 17-18.

Pujar, N.S. and Savanurmath, C.J., 1998, Innovations in Indoor maintenance of tasar silk moth, *A. mylitta* Drury. *Proc. III*[rd] *Int. Conf. on wild silkmoth*, p.p. 61-65.

Purushotham Rao, A., 1998. Some Salient features of Andhra local ecorace, *Antheraea mylitta* Drury in relation to its conservation and multiplication. *Proc. IIIrd Int. Conf. on wild silkmoth*, p.p. 356-358.

Puttaraju, H. P and Nagaraju J., 1998. Preliminary observations of B-chromosome in the silkworm, *A.roylei* (Lepidoptera – Saturniidae). *Current Science*. 54: 471 – 472.

Rajadurai, S. and Thangavelu. K., 1998. Biology of Moon Moth *Actias selene* Hubner (Lepidoptera: Satumiidae) prevent in Bhandara forest, Maharashtra. *Proc. IIIrd Int. Conf. on wild silkmoth*, p.p. 362-366.

Rajany, T., Mathew P.J., Remadevi O.K and Srivastav A. K. 2005. Prospects of tasar culture in Kerla. Advance in tropical sericulture. (edit. Dandin and Gupta) Pp.422 – 425.

Ramanathan, A. 1997. Magic of silk, *Indian silk* 36(9): 6-10.

Rath, B. B. 1969. A study on double cocoons spun by tasar silkworm. *Indian silk* 8(4): 7.

Rath, S.S., 1998, Studies on growth and development of *A. mylitta* Drury (Lepidoptera–Saturnidae) fed on three different natural host plants. *Proc. IIIrd Int. Conf. on wild silkmoth*, p.p. 99-103.

Reddy, J. K., Jayaprakash P., Yadav G.S., Seshubabu, V and Sinha B.R.R.P. 1998. Utilization of Andhra and Bhandara local ecorace for prospective production of tasar cocoons. *Proc. IIIrd International conference of wild silkmoths*. Pp. 348-355.

Roo, K.V.S., Ramkumar, Mahobia G.P., Pande, K.V., Ade, A.L., and Sinha, B.R.R.P., 1998. Studies on it situ conservation of Raily ecorace of Indian wild Tasar insect *A. mylitta* D. (Lepidoptera–Saturniidae). *Proc. IIIrd Int. Conf. on wild silkmoth*, p.p. 314-316.

Saikia, B and Handique R. 1998. Biology of a wild silk moth, *Attacus atlas* L. *Proc. IIIrd Int. Conf. on wild silkmoth*, p.p. 345-347.

Sathe, T. V. 2006. Sericulture can prevent soil erosion and deforestion. Environmental issue and options (Edt. Mishra *et al.*), 22, Pp 453-465.

Sathe, T. V. 2007. Biodiversity of wild silkmoths from Western Maharashtra, India. *Bull. Ind. Acad. Seri.* Vol. 11(1). Pp.21-24.

Sathe, T. V. and Jadhav, A. D. 2001. Sericulture and pest management, *DPH New Delhi*, Pp.1-197.

Sathe, T. V., Mulla.M. K. and D. B. Sathe. 1997. A brief note on the silkworm diversity in western Maharashtra. *Proc.8th mani. sci. congr. Biodiversity and resource management*, pp.7-8.

Sathe, T.V and Mulla, M.K. 1999. Insect pests of mulberry from Kolhapur, India.. *Geobios* new reports, 18, 73-74.

Sathe, T.V and Pandharbale, A.R 1999. Halk moth (Spingidae: Lepidoptera) diversity in Western Maharashtra including Ghats. *Geobios*.7, 77-82.

Sathe, T.V and Pandharbale, A.R 2004. Biodiversity of moths from Western Ghats of Satara districts, Maharashtra. *Bull. Bio. Sci.* pp. 77-82

Sathe, T.V and Pandharbale, A.R 2008. Forest Pest Lepidoptera. DPH, New Delhi.Pp.186.

Sathe, T.V and Thite, S. H. 2004. Shoot feeding and sericultural trends. DPH Delhi.pp.1-176.

Sathe, T.V. 1998. Sericultural crop protection. Asawari pub. Os'bad. pp. 1-90.

Sathe, T.V., Jadhav, A. D., Kamdi, and J.P. Undale. 2008. Low cost rearing technique for mulberry silkworm (PM × NB_4D_2) by using nylon and indigenous shelves. Biotechnological approaches in entomology. 5, 205-211.

Sathyanarayan, K. and J. V.K. Rao (2004) Central Silk Board Bangalore. *Indian Silk*, 43(2), 23–26.

Seitz, 1913. Macro Lepidoptera world (1933), India. Vol, 10 pp- 509-516.

Sengupta, A. K., Sinha. A. K and Sengupta, K. 1993. Gene' reserves of *A. mylitta* Drury, *Indian Silk*. 32 (5): 39-40.

Sengupta, K. 1987. Current status of non mulberry sericulture and its future development. *Sericologia*, 27(3): 475-480.

Sengupta, K. Simbli, S.N., Saddurin and Singh, M.K. 1986. High altitude cocoon preservation and general grainage operation Ann. Rep. Central tasar research, Ranchi, India, p.p. 396-399.

Shamitha, G and A. Purushotham Rao. 2002. Problems and prospects in indoor rearing of tasar silkworm. Advance in Indian sericulture research (edit. Dandin and Gupta). 409–411.

Shamitha, G. 2007 Total indoor rearing of the tasar silkworm. *Everyman's Science*, Vol. XLII-(4): 198-202.

Shankar Rao, K.V., G. P. Mahobia., Roy. G. C and Sinha B.R.R.P.2004. Ethology and conservation of Raily ecorace of Indian wild tasar insect *A.mylitta* D. *Ind J. Seric*..43 (1): pp. 71-77.

Shankar Rao, K.V., Ramkumar., Roy G.C and Sinha B.R.R.P. 2002. Biology of raily silkmoth, *A.mylitta* D.(Lepidoptera – Saturniidae). *Proc XIX*[th] *congress of the international sericultural commission*. Thailand. Pp.279 – 284.

Shankar Rao, K.V.S., Mahobia G. P., Ramkumar., Roy C.G. and Sinha B.R.R.P.2005.Reproductive potential of raily tasar silkworm A.mylitta D. under natural and captive conditions. Advance in tropical sericulture. (edit. Dandin and Gupta) Pp.393–396.

Shetty, K. K and Samson M. V., 1998. Non – Mulberry sericulture in India. *Indian silk*, 37 (6 and 7): 21 – 25.

Shi, J., D. G. Heckel and M. R. Goldsmith. 1995. A genetic linkage maps for domesticated silkworm, *Bombyx mori*, based on restriction fragment length polymorphisms. *Gent.Res*.66. 109 – 126.

Singh, B. M. K. and Srivastava, A K. 1997. Eco-races of *Antheraea mylitta* and exploitation strategy through hybridization. Current *technology seminar on non-mulberry sericulture*, conducted by Central Tasar Research and Training Institute, Ranchi–8, 9 Jan, 1997.

Sinha, A. K. Choudhary, A. and Reddy, K. 1994. Artificial mating: A tool for additional seed production in tropical tasar. *Indian silk*, 32 (8): 37-39.

Sinha, B. R. R. Pd. 1998. Ecoraces of Indian wild Silkmoths. *Indian Silk*. 37 (6 and 7): 38.

Sinha, B. R. R. Pd., 1998.wild silk research in India: Present and future. *Indian silk*, 37 (677): 27- 31.

Sinha, R. K. Kalashreshtha V., Mishra, P.K. and Thangavelu. K. 1992. Constraints of the tasar industry in India. *Indian silk*, 3(1): 31.

Sinha, U.S.P., Bajpeyi, C.M., Sinha. A.K., Brahmachari, B.N. and Sinha, B.R.R.P. 1998. Food consumption and utilisation in *A. mylitta* Drury larvae. *Proc. IIIrd Int. Conf. on wild silkmoth*, p.p. 182-186.

Slansky, F. Jr and Scriber J. M.1984. Food consumption and utilization incomprehensive insect physiology, Biochemistry and pharmacology. (Edt.G. A. Kerkat and L. I. Gilbert), 4. 87–163.

Sonwalkar, T. N. 1991. Hand book of silk technology. Wiley eastern Ltd. New Delhi. 58–98.

Sonwalkar, T. N., 1986. Silk reeling technology in India, Lectures on sericulture, Bangalore, p.p. 205-219.

Sonwalkar, T.N. and Jolly, M.S. 1985. Investigations on the influence of colour a site of Daba cocoons on reeling performance. *Ind. J. Seric.*, 24(1): pp.1-6.

Srivastava, A.D. and Ansari, M.O.A., 1979. Demonstration of new technique of cooking and reeling of tasar cocoons. Ann. Rep. Central Tasar research Ranchi. India. p.p. 84.

Srivastava, A.K., Nagvi. A. H., Roy. G.C. and Sinha, B.R.R.P., 1998, Temporal Variation in qualitative and Quantitative characters of *A. mylitta* Drury. *Proc. IIIrd Int. Conf. on wild silkmoth*, p.p. 54-56.

Sudhakar Babu, C. H. and Purushotham Rao A., 1998. Indoor Chawki rearing trey–A new device for tasar silkworm *A. mylitta* Drury. *Proc. IIIrd Int. Conf. on wild silkmoth*, p.p. 87-91.

Tanaka, Y., 1950a. Studies on hibernation with special reference to photoperiodicity and breeding of Chinese tasar silkworm. *J.Seric.Sci.Jap*, 19 (4): 358.

Tanaka, Y., 1950b. Studies on hibernation with special reference to photoperiodicity and breeding of Chinese tasar silkworm. *J. Seric. Sci. Jap*, 19 (4): 429.

Tanaka, Y., 1950c. Studies on hibernation with special reference to photoperiodicity and breeding of Chinese tasar silkworm. *J. Seric. Sci. Jap*, 19 (4): 520.

Tanaka, Y., 1956. Photoperiodic effect of alternated dark and light phase of diapause of *A. pernyi*. Ann.Rep. Natl. Inst. Genetics, 6: 35-36.

Thakur, S.S., 1994. Present studies and scope of Indian tasar culture 2nd.*Int.conf.on wild silkmoth*, Aug.18-22, Hotaka, Japan.

Thangavelu, K and Sahu A. K. 1986. Further studies on the Indian rearing of muga silkworm *A. assama.Sericologia*, 25(2-3): 153-158.

Thangavelu, K. 1992. Population ecology of *A. mylitta* Drury (Lepidoptera–Saturniidae). *Wild Silkmoths*, p.p. 99-104.

Thangavelu, K. 1993. Identification of problems and prospects in tasar culture. CST UTI, Ranchi. Proceeding o the workshop held on 29-30, Aug 92, Kakatia University and C.S.B., Warangal. India. p.p. 57-62.

Thangavelu, K. and Sinha, A. K.1993. Population ecology *A. mylitta* Drury (Lepi: Satur). *Wild Silkmoth*, 92. 88-92.

Thangavelu, K., 1991. Problems and prospects of tasar culture in India. *Sericologia*, 31 (1): 205-208.

Thangavelu, K., 1991.Wild sericigenous insects of India.A need for conservation. *Wild silkmoths*. 91, 1992. 71-77.

Thangavelu, K., 1992. Recent studies in Indian tasar and other wild silkmoths. *Wild Silkmoths*, 91, 20-29.

Thangavelu, K., Bajpayi, C. M. and H. R. Bania.1991. Indoor rearing of Tasar silkworm diet. *Indian Silk*. 30 (6): 19-20.

Thangavelu, K., Bhagwati A.K. and Chakraborty, A.K., 1987. Studies on some wild sericigenous insects of North-eastern india. *Sericologia*, 27-1. 91-98.

Tiken singh, N and B. B. Bindroo., 2006. Ideal rearing sites for oak tasar culture in north – western Himalayas. *Indian silk*, 45(6): 14-16.

Tiwari, S. K., 1985. Seri. Ecosystem requires protection with special reference in Indian non-mulberry silk industry. *Sericologia*, 25 (1): 115.

Tiwari, S. K., 1997. Arjun plantation under social forestry. *Indian silk*, 36 (6): 33-37.

Tiwari, S.K., 1998. Tiwari grainage tray: a break through in tasar seed preparation. *Indian silk*, 30(12): 18-22.

Tutt, 1899-90, British 7 Lep. Vol-8.

Waldbauer, G. P. 1968. The consumption and utilization of food by insect. Advances of insect physiology, 5. 229 – 288.

Watson, J. H. 1911. The wild silkmoths of the world with special reference to the Saturniidae. Manchester, School of technology, Manchester, England. Pp.8.

Zhuang, Da Huan, Liu Shi xian and Li Long.1994. Sericultural production strategies in the 21st Century. Paper presented at International Conference on Sericulture. *Indian Silks*, 33 (8): 4-9.

Index

A

A. alias 58

A. assamensis 59

A. atlas 6, 39, 59, 60, 122, 123, 126, 128, 133, 136, 138, 143, 184

A. cingalesa 60

A. edwardsi 58, 60

A. mylitta Drury 52, 198

A. paphia 33, 37, 60, 194

A. pernyi 10, 15, 187

A. roylei 15, 59

A. selene 6, 122, 127, 134, 136, 138, 144, 147, 150, 153, 185, 186, 197, 198

Abdomen 68

Actias 60

Actias callandra 185

Actias indica 60

Anacardia occidentals 194

Anaphe moloneyi 59

Anaphe panda Bosid 2

Animal kingdom 55

Antheraea assama Ww 1, 15, 40

Antheraea hubner 73

Antheraea mylitta 1, 8, 13, 19, 33, 37, 39, 40, 59, 60, 72, 123, 187, 190, 195

Antheraea mylitta arjuni sub. sp. nov 73, 84, 197

Antheraea mylitta badami sub. sp. nov 74, 109, 197

Antheraea mylitta grayi sub. sp. nov 74, 89, 197

Antheraea mylitta indica sub. sp. nov 73, 74, 84, 197

Antheraea mylitta jujubi sub. sp. nov 73, 79, 197

Antheraea mylitta kolhapurensis sub. sp. nov 74, 94, 99, 104, 108, 123, 136, 138, 162, 187, 195, 197, 198

Antheraea mylitta koynei sub. sp. nov 74, 99, 197

Antheraea mylitta sahydricus sub. sp. nov 74, 113, 114, 118, 197

Antheraea mylitta sathei sub. sp. *nov* 74, 104, 108, 1097

Antheraea mylitta sub sp. nov *kolhapuriensis* 128, 197

Antheraea pernyi G.M 1

Antheraea proylei 59, 193

Antheraea yamamai G.M 1, 10, 59, 187, 190

Attacus 60

Attacus atlas L. 125, 197, 19?

Attacus synthia 11

Avicennia alba Bl. 133

B

B. mandarina M 2

Biodiversity 12, 17, 37

Biology 121

Biology of *A. mylitta* sub sp.nov *kolhapuriensis* 124

Biology of *A.mylitta* sub sp.nov *kolhapuriensis* 124

Biology of *Actias selene* (Hubner) 124

Biology of *Attacus atlas* (Linnaeus) 123

Biology of moon moth 39

Biotechnological approaches 7

Bombyx mori L. 2, 16, 186

Bruguiera gymnorhiza (L.) 134

Butterflies 56

C

Cedrela paniculata 134

Circadian behaviours 11

Cocoon 81, 126, 138

Cocooning 193

Collection 53, 54

Coriana nepalensis 134

Cornus officinalis Sieh 134

Cricula 60

Cricula trifenestrata Helfer 2, 59

Cryptothelea formosicola Strand 2

D

DNA markers 9

E

Economic Plantation 22

Ecoraces 8, 13

Ericulture 5

Etymology 77, 83

F

Fabrication 23

Fecundity 194

Ficus carica 46, 123

Food Consumption 145

Food Energy Assimilation 147

G

Genitalia 51, 61, 68, 76, 80
Gonometa postica Walkar 2

H

Heptapleurum octophyllum 134
Heteropanax fragrans Seem 16
Holotype 77, 83
Host Plants 77, 83
Hybrid 198
Hybridization 9

I

Indoor rearing 190, 192, 195

L

Laepa katinka 60
Lannea coromandelica 185
Lemaireia 60
Lepidoptera 56
Life cycle 197
Litsea polyantha Juss 17, 20
Loepa 60

M

Machilus bombycina King 17
Manihot utilissima Pohl 17
Meyna laxiflora 133
Morphological 61
Moths 56
Moyna laxiflora 122

N

Notolophus sp 22
Nursery Technique 21

O

Orders 55
Outdoor rearing 195

P

Pachypasa otus 59
Parasites 42
Paratype 77, 83
Philosamia cynthia D 10
Philosamia ricini 16, 20
Photography 52
Photoperiodism 40
Polyphagous 8, 191
Predators 42
Preservation 53

Q

Quercus semicarpifolia 193

R

Rearing 52
Reeling 23, 32
Respiration 147
Review 37
Rhodinia 60
Ricinus communis 3, 17, 20, 189

S

Samia cynthia ricini Hutt 1, 59
Saturnia 60
Saturnia anna 60
Shorea robusta 11, 35, 191
Silk textiles 1
Solus 60

Sphenoptera sps. 22
Survey and Surveillance 53

T

T. arjuna (Arjun) 5, 11, 30, 35, 46, 52, 77, 83, 88, 95, 96, 102, 107, 112, 117, 123, 124, 128, 134, 147, 185, 197
T. catappa 46, 52, 77, 83, 96, 128, 153, 194, 197, 198
T. tomentosa (Asan) 11, 35, 46, 52, 77, 83, 88, 95, 96, 107, 117, 123, 124, 128, 134, 150, 185
Tasar culture 41
Tasar sericulture 196
Tasar silk 137
Taxonomical 38
Taxonomy 54, 55, 197
Trioza fletcheri Minor 22

V

Voltinism 39

W

Wild Silk Export 27
Wild silkmoths 197
Wings 68

Z

Z. alatum 134
Z. jujuba 46, 77, 83, 88, 93, 96, 102, 107, 112, 117, 124
Z. mauritiana 134, 185
Zanthoxylum acanthopodlum 134
Zizyphus 20